A Hand to the Plough

a farmer's vision for the
twenty-first century

Patrick Evans

SAPEY PRESS

Published in 2006 by the Sapey Press, Whitbourne, Worcester WR6 5ST

Copyright © Patrick Evans, 2006

The right of Patrick Evans to be identified as the author of this work has been asserted by him in accordance with the Copyright, Designs and Patents Act 1988.

All rights reserved. Except for brief quotations in a review, this book, or any part thereof, may not be reproduced, stored in or introduced into a retrieval system, or transmitted, in any form or by any means, electronic, mechanical, photocopying, recording or otherwise, without the prior written permission of the publisher.

ISBN 0 9528916 3 8

A full CIP record for this book is available from the British Library
A full CIP record for this book is available from the Library of Congress
Library of Congress catalog card: available

Cover designed by Paula Larsson
Text designed and typeset in Minion by Dexter Haven Associates Ltd, London
Printed and bound in Great Britain by William Clowes Ltd, Beccles, Suffolk

A Hand to the Plough *is a remarkable compilation of inspiring essays which show the path to sustainable human happiness and well-being. The last chapter on the unity of the spirit brings out the great importance of integrating science and spirituality. This book is a must for all interested in fostering harmony with nature and with each other.*

Professor M.S. Swaminathan

> *Chairman, National Commission on Farmers, Government of India*
> *President, Pugwash Conferences on Science and World Affairs*
> *Chairman, M.S. Swaminathan Research Foundation, Chennai, India*

By the same author
Farming For Ever, Sapey Press

A Hand to the Plough

To Peter Howard, a friend and a great man

CONTENTS

Preface		ix
Part I: A twenty-first-century perspective		1
1	The evolution of humanity	3
2	Biotechnology and the onward march of science	10
3	Biotechnology in agriculture	22
4	A shift in economic thinking	37
5	Economic structures and the meeting of need	46
6	Towards an international perspective	59
7	The political dimension	66
8	The current political challenge	73
9	Religion – the common heritage of mankind	82
10	God is His own interpreter	94
Part II: The farmer's contribution		105
11	Why farming is different	107
12	The Common Agricultural Policy (EU) – the way forward	116
13	Europe – the enlarging union	129
14	Developing patterns in agriculture – Asia	138
15	Developing patterns in agriculture – Africa	151
16	Developing patterns in agriculture – Latin America	160
17	The twin poles of globalisation	168
18	Growing character as a crop	175
19	The International Farmers' Dialogue	184
20	The unity of the spirit	195
Bibliography		201
Index		203

Preface

I am conscious in trying to write this book that I come from a Western background, and cannot be objectively universal. On the other hand, if globalisation means anything, it means attempting to take a world view about what lies ahead. That may sound excessively ambitious, but it must be attempted if humanity is to move on to a new stage in history. Inevitably I shall be right about some things and wrong about others, even though I set out to serve the truth. But it will be a voyage of discovery which I hope I can successfully share with others, who will prove themselves better navigators.

In making the attempt I am influenced by Lord Selborne's comment on the need to engage with others. 'There has to be a culture of openness, transparency and inclusiveness. Our future as an industry will depend on our success in communicating more effectively with the interested observer of the agricultural scene.' I am sure that is true, and I also hope that it may be possible to raise a much wider interest in what agriculture has to offer. It does in fact look very different from the other side of the world.

In expressing thanks for the help I have received, my first debt is to the Revd Robin Denniston and Alec Porter, who have read the earliest draft as it was written. They have been diligent commentators. Then to Nick Wallis, particularly for advice on the economic chapters. Although a tropical agriculturist disclaiming any special expertise in economics, he spent a large part of his career with the World Bank. And to John Hodges, Editor of *Livestock Production Science*, the journal of the European

Association for Animal Production. He has always been committed to making ethics a priority in science, and this concern is also reflected by many other contributors to the journal.

A number of others from farming and other backgrounds have also read individual chapters or discussed ideas. It may not be an exhaustive list, but I am grateful to Claude Bourdin, Charles Danguy, Matt Dempsey, Amina Dikedi, Puntipa Pongpiachan, Luis Puig, Dr Ian Robertson, Peter Rundell, Professor M.S. Swaminathan, Laurie Vogel and Jim Wigan. Also my brother Robin. Finally a special thank you to my wife, Kristin, who is the mastermind on the computer, and an ever-present help in bringing some of my flightier ideas down to earth.

I have also included substantial material from articles appearing in the magazine *For A Change*. Its mission statement reads: '*For A Change* is about change, how to make it happen and how to live it. We believe that what happens inside people has an effect on the world around them. At the heart of global change lies change in the human heart. We draw our material from a wide range of sources, including Initiatives of Change. We give a voice to people all over the world who are making a difference. We invite our readers to join them. Your stories are our stories.'

Michael Smith, whose *Beyond the Bottom Line* is also quoted, is Associate Editor of *For A Change* (published six times a year by Initiatives of Change, 34 Greencoat Place, London SW1P 1RD, UK).

For the production of the book and all the coordinating work involved, I am grateful to Robert Hastings of Dexter Haven Associates Ltd. He has always been available, and instantly ready to see things through.

Pat Evans
Whitbourne

PART I

A twenty-first-century perspective

1

The evolution of humanity

The book of Genesis, which is the first book in the Bible, opens with the statement that 'the Earth was without form and void'. It could be taken as a modern metaphor for the moment of the Big Bang. For what is the Big Bang but man's effort to explain something which is still beyond his understanding? It remains a source of wonder.

Human evolution in many ways is in its infancy. The suggestion by Francis Fukuyama that we might be in the final phase of human history was not only premature but positively rash. The growing debate on climate change, and the wider issues of environmental practice, open up a new historical perspective. Man's pursuit of an affluent lifestyle, combined with a vast increase in world population, is raising questions on the sustainability of our present activities. Science has provided us with the means to reach our present position, but has not perhaps generated adequate thought to the ends on which we should focus.

The real point is that the evolution of mankind has brought a whole new element into play. The development of people is at root a spiritual question, and not just a question of competition.

It has introduced the concept of right and wrong as the arbiter of further progress rather than a continuing struggle for existence. Religion in the sense of a relationship with God began with perceptions of life revealed in nature, and has been moving forward ever since. It has brought another realm into the equation with an urgent need to measure new challenges afresh. Instead of closing down historical perspectives, it may be offering a new vista which could take centuries to make our own.

Julian Huxley, the zoologist, wrote in 1957: 'It's as if Man had been suddenly appointed Managing Director of the biggest business of all – the business of evolution. Whether he wants it or not, he is determining the direction of evolution on the earth.' Such a statement carries several implications. First, perhaps, that it is possible mankind represents the ultimate in the evolutionary process. If this is so, we may need to go much deeper into discoveries of the spirit. It has been suggested that there are now no pressures on people to improve physically, because modern aids can support failing eyesight and hearing, while competitive progress depends more on brain than brawn. If large numbers of people are destined to be glued to their computers, physical development will depend on those who are dedicated to an increasing variety of sports. Already obesity has become an issue, and whatever science fiction may be written about super-intelligent computers of the future, they will first have to be created by super-intelligent people.

If all this should come to pass, people cannot shoulder such new responsibilities without some better understanding of the purposes of Creation. Darwin made a huge contribution to our knowledge with *The Origin of Species*, but he never made the claims some have attributed to him. As Canon B.H. Streeter, the Oxford theologian, put it, 'What Darwin had discovered was neither the nature of life nor the goal of its endeavour, but the road by which it had travelled'.

The evolution of human society nevertheless urgently needs pursuing. Some may feel that progress has been slow, and that past civilisations are witness to man's inability to sustain it. Yet in the aeons of evolution, man's history is extremely short, and in broad terms there has already been a considerable progress in human attitudes. We have by no means banished wars, but using violence to attain national aims has become unacceptable to the international community. On average there are more than one hundred armed conflicts in the world in any given year. So it is not the diminution in the resort to war as such, so much as a growing conviction in public opinion. This revulsion is being further reinforced by the mindless violence of terrorism.

Economics has taken over as the motor of human ambitions, but even here questions are being asked about the nature of competition and the way money is handled. The power of huge multinational companies is increasingly seen to need the restraining hand of government, but democracy may increasingly require completely new ways and procedures. As Amartya Sen (1998 Nobel Prize-winner in economics) has observed of globalisation, 'The debate is rather about inequality of power, for which there is much less tolerance now, than in the world which emerged at the end of the Second World War'.

If power no longer 'comes out of the barrel of a gun', in Mao's familiar phrase, it must also not come from the depth of a bank balance or the capacity to 'shock and awe' with missiles, a phrase which may come to haunt its authors. That puts the accent once more on human values, and returns us to the distinction between man and other animals. Professor Henry Drummond of Edinburgh, who was both a scientist and a Christian, pointed out that the creation of our planet was no less a miracle because it had taken place over billions of years. He speculated that if evolution is simply a method of creation, why was it chosen? It must have been foreseen that the time would come when the directing of part of the course of evolution would pass into the hands of man.

For Drummond, who grew up in the wake of the stunning impact made by Darwin's book, the key issue was that the struggle for existence had two strands. One was 'the struggle for life' (finding food), the second, the struggle for the life of others (reproduction). He considered that altruism (or unselfishness) in the higher animals is the direct outcome of the reproductive process, which in mammals came to be associated with a passionate, protective instinct. 'To affirm that altruism is the peculiar product of religion is to excommunicate nature from the moral order, and religion from the natural order.'

For Drummond also, the evolution of love is a piece of pure science. The struggle for life is for individual survival. The object of reproduction is to secure the life of the species. Both are selfish at the outset, but there is quite early a parting of the ways. For love in its true sense has to become self-sacrifice. It is in the care and nurture of the young that altruism finds its first expression. Both literally and scientifically, love is life. For sacrifice is always involved in the continuance of life, and co-operation is a gift arising out of it. It is clear too that human children take longer to reach maturity than other mammals. They need care over a sustained period of years.

Looking at the next stages in evolution, Drummond builds on the idea that the struggle for the life of others has become wholly a social force. He mentions the conclusion of some that man is not only the highest branch of Creation, but the highest possible branch.

> So man has entered into his kingdom. As mental evolution succeeds organic, man stands alone in the foreground, and a new thing, spirit, stirs within him. Evolution is not a progress in matter, because matter cannot progress. It is a progress in life and spirit, in that

which is limitless, in that which is at once most human, most rational and most divine. Science doesn't actually know what these forces are. It only describes and classifies them.

This brings up the questions raised by the evolution of human nature, and what we should really be looking for in the twenty-first century. When William Penn said: 'Men must choose to be governed by God or they condemn themselves to be ruled by tyrants,' he uttered a truth which still holds good. Quite apart from questions of faith and religion, it is an essential challenge for the future of democratic government. For whatever shape or form democracy may take, it is increasingly being seen as the way ahead worldwide. Yet it cannot work effectively without self-discipline and the acceptance of an authority greater than the State, which must be defended by individual conscience. That – much more than casting a vote – should be the true commitment of every democrat.

Above all it underlines the need to go beyond talking to living. This is essential because experience is the laboratory of life, and whatever our faith or lack of it, there is a need to see our ideas put to the personal test. Those who enter the laboratory in the pursuit of truth will not shrink from the most rigorous examination. They will not attempt to give their own ideas a fair wind by rigging whatever surrounding circumstances they may be able to influence. Because the first test is our own fitness to participate, and what truth means in our own living. As Mahatma Gandhi put it, 'Be the change you want to see'.

In speeches made in the United States in the last year of his life, Peter Howard, a world leader of Moral Re-Armament, touched a number of times on the next stage in the evolution of humanity. He paid generous tribute to all that America has given the world, but he could see the need for huge changes. He was crystal clear in his moral challenge, but he also had the faith to expect a visionary leap forward on the issues to which we may dedicate our lives. 'We need a revolution to carry the whole world forward fast to its next stage of human evolution – to outpace the growth of human power, wealth and skill with a growth in human character.'

Looking back in history, it is possible to detect a development in human attitudes, even if it can hardly be described as a seamless progress. Jesus Christ is variously regarded by many people today as a rebel or a non-violent revolutionary, but few have measured the extent of the crop still growing from the seed he planted. Yet Alexandre Dumas (fils) could write more than a century back: 'The world is entering an epoch when the words "love one another" will be accepted without arguments about whether it was a man or a God who spoke them'. It is a paradox of the times in which we live that current horrors may not erase this comment. But they certainly underline the need for a quantum leap forward.

It is a big question whether powerfully entrenched interests can be stirred to take a new direction. But if the growth of democracy round the world is to mean anything, it must find ways to go beyond protest and criticism to the attractions of such a fresh direction. It has been widely observed in the field of international development, that initiatives from the bottom up are the most practical and have the best chance of success. So that is, in itself, an endorsement of the democratic way. Current trends in globalisation would seem to imply that economic power should be concentrated in larger and larger organisations where policy is under central direction. But the limitations of such arrangements are beginning to be more and more apparent. Control of policy is being increasingly decentralised, and even then there may be a better future in partnerships than large-scale control, where devolution has to be organised rather than developed.

THE EVOLUTION OF HUMANITY

This book is an attempt to review the developments of one lifetime, and to focus the direction in which human destiny beckons. It may seem an ambitious undertaking, but it is in truth more an examination of how the ordinary man can still have an influence he may have thought was denied him. The complications of life do no more than underline our need of teamwork in tackling them. They do not imply that it is beyond us to understand the basis of the trust required, or the motivation needed to move in the desired direction.

2

Biotechnology and the onward march of science

In his book *The Darwin Wars*, Andrew Brown goes so far as to talk about the scientific battle for the soul of man. He suggests that in the last thirty years the resurgence of Darwinian explanations has provided a particularly potent brew of good science, striking metaphors and bad philosophy. Yet he concludes, 'Darwinian beliefs have been used to justify a variety of causes. But all the justifications are agreed on one thing – that there is a human nature, and that the study of our evolution can help us to discover how we ought to live'.

Andrew Brown further suggests that Christianity has done a much better job (than science) of assimilating its insights into human nature to shape rules of conduct, which can stand independently of the beliefs which have shaped them. He writes,

> Moral philosophy is what has made most of the fuss because it is much more interesting and important than science. Most of the world's population have lived, and do live to the best of their abilities, innocent of science. Everybody does moral philosophy every day, if only on a practical level.

Nevertheless science, even on an academic level, postulates a direction in evolution. Julian Huxley, the zoologist, wrote in 1926, 'We cannot say that Evolution is purposeful until we are privileged to know what processes occur in the thought of God: but we can and must say that it has direction', and that direction has to be expressed in the development of life itself. With the advent of man it becomes a question of conscious behaviour. The popular misconception that we are entirely programmed by our genes seems to rest on the assumption that there is some general law of progress in biological evolution. But while our genetic makeup may give us physical and mental strengths or weaknesses, behaviour itself is the property of living things. Genes themselves can have no obligatory influence on the way we choose to behave, and the whole elucidation of their control and influence still lies in the future.

Evelyn Fox-Keller (*The Century of the Gene*) points out that there is a huge gap between genetic information and biological meaning. She writes, 'What is most impressive to me is not so much the ways in which the Genome project has fulfilled our expectations, but the way in which it has transformed them'. Now we have to move from structural to functional genomics, because the sequence is not an end product but a tool. It is, for example, already enabling DNA to become an important element in crime detection. While the fact that DNA is the same chemically for people, animals and plants would seem to show that the nature of life is one and indivisible, it is a surprising fact, given the diversity of its expression.

> With the advent of the recombinant DNA revolution, it is possible to insert genes from one species into another without upsetting genetic stability. In fact the evidence of robustness in cellular systems means they are probably impervious to single gene changes. Yet it is the success of genetic engineering which has convinced

> molecular biologists of the causal power of genes. 'Cause' may be all right in a limited pragmatic sense covering practical results, but it does not embrace the whole process or long term consequences. So demonstrations of efficacy can offer little assurance to those who worry about these other questions.

This underlines again that genes determine possibilities, covering a range of behaviour, but humans in their interaction with the varied pressures of the environment make choices. So it is the act of choosing which becomes the new factor in evolution.

Fundamentally the processes of life are a great deal more complicated than was supposed even thirty years ago. It may be that the mechanisms of life cannot be dissected beyond a certain point, and that only the organism itself, as completely put together, is able to function. It suggests to many that the primacy of the gene as a core explanatory concept is more a feature of the twentieth century than it will be of the twenty-first. For a physicist biology is not governed by laws, in the sense that there are no immutable rules which dictate certain consequences. Life as we know it is the beneficiary of a long history of fortuitous opportunities. Yet amid all these possible chance permutations the ability to stay on track is the key to biological development. Computer technology owes its advance to the changing models of human design, while organisms evolve without the benefit of a human designer. That is the elusive nature of life, which goes far beyond the imaginings of robotics or science fiction.

But one cannot quite leave it at that. Evelyn Fox-Keller writes,

> Darwin taught us the importance of chance in evolution by natural selection, but he also taught us the importance of challenge. Thus I prefer to end with the prediction of a great deal more to come, perhaps even of another Cambrian period, only this time not in the realms of new forms of biological life, but in new forms of biological thought.

Commenting on the success in unravelling the human genome, she takes comfort in the thought that 'it is a rare and wonderful moment when success teaches us humility'.

Pure science as the pursuit of truth should always point towards humility, because the more we know about a subject, the more there is to know. Moreover, the scope of each subject extends far beyond one person's understanding. Scientists have to trust other people's work, because they are constantly building on work they have not done, and may not even fully understand. That is partly why cheating in science is such a heinous offence. A false experiment may confuse the work of many others. Generally speaking this is covered by results only being accepted when they have been successfully replicated, but besides such insurance it may be useful to have a nose for false scents.

The layman may find that the more controversial area is where science and commerce meet. In the realm of genetically modified organisms, Monsanto may have seriously overplayed its hand, but when the language of scientific development works well, the research scientist comes to have his own investment in its rhetorical power. So it pays to be cautious, and to realise that the consequences of many new discoveries trumpeted in the Press will only be putting in an appearance many years later. But that will in no way inhibit the heated discussions of possible danger, and for the moment it has unfortunately become a dialogue of the deaf.

Much is now made of the Precautionary Principle, which postulates the need to take measures to protect the environment or human health where there are possibilities of serious or irreversible damage, although scientific evidence may be lacking. Even its keenest advocates are inclined to describe it as a rather shambolic concept, but it seems that the gap between public perception of risk and 'expert' assessment of risk is getting wider. Talk of sound science is heavily discounted, yet Greenpeace is credited with convincing the leadership of Royal Dutch Shell that a new company policy on the environment had to be developed. It is ironic that in the case for the disposal of the Brent Spar oil rig which occasioned it, Greenpeace was actually mistaken scientifically. Shell had opted for sinking the platform in the deep ocean, until Greenpeace forced the case for bringing it to break up on land in Norway. Yet scientific investigation of the choices showed that the former course would have been environmentally preferable. It was a demonstration that the Green lobby has proven public relations and marketing skills, and both Shell and BP have now taken a leaf out of their book.

Trust is all-important in dealing with risk. The logical end of the Precautionary Principle would be that we will try nothing new, and so never make progress. Public trust was lost over the government's handling of BSE and Foot and Mouth, which greater scientific input might have prevented. Although it has to be said that the public do not like to be told that scientists are still in the dark. There should also be more scepticism about what appears in the media. The origin of BSE in cows remains uncertain, though there is virtually no doubt that vCJD is caused by BSE prions gaining access to humans who are genetically susceptible. Whatever their origin in cows, if meat and bonemeal had not been fed, the disease would not have spread to other cattle. It would simply have remained an unidentified cause of death in a small number.

Feeding meat and bone meal was introduced more than seventy years ago, primarily as an efficient way of disposing of offal, and providing a new cheap source of animal protein for livestock. It appeared to be both benign and cost-effective, but there was a failure to engage in wider consultation. The decision was made by scientists, veterinary and business interests with government approval. The veterinary and government interest was limited to ensuring that pathogens were eliminated, but abnormal prions appear able to survive sterilisation. Wider consultation in the food chain might have stopped the process in its tracks, since it is clear that today's public is shocked at the cannibalism of feeding animal protein to vegetarian ruminants. It is certainly true that today, ethical considerations must be part of such decisions, and must embrace all those involved in the food chain, including consumers.

For although it is true to say that genetically modified plants have entered the food chain without ill effect, there have been particular cases of failure. One was Star Link corn (maize) in the United States, which passed all its tests, but then had to be withdrawn because it triggered allergies. Another was Tryptophan, an amino acid found in many foods (notably in milk and bananas), and commonly manufactured for inclusion in popular nutritional supplements. It may be worth recounting the experience briefly, since it occurred only in the United States, where a contaminated source appeared.

Tryptophan is essential to the human body for the production of serotonin, a brain chemical necessary for sleep and for mood regulation. It is also the nutrient which the body uses to make melatonin. During the 1980s it was widely available as a supplement in health food stores, supermarkets and drugstores. In November 1990 it was taken off the market by the Food and Drug Administration (FDA), because of its statistical connection to a dangerous blood disease Eosinophilia Myalgia Syndrome (EMS). Between July 1989 and December 1990 more than 1500 cases of EMS resulted in 27 deaths. A contaminant was found in Tryptophan manufactured by Showa Denko KK, one of six Japanese companies providing nearly all the Tryptophan used in the United States. The contaminant appeared when at the end of 1988 the production process was altered, to allow a reduction in the amount of activated charcoal purification by using genetically modified bacteria. It is apparently unclear whether the new strain of bacteria or the reduction in charcoal purification was responsible for the EMS-producing contaminant. It was identified as an accidentally synthesised new amino acid, described as Ethylidenebis Tryptophan (EBT). Research into this new amino acid is continuing, but no problems have been encountered from any other source.

All of which underlines the need for vigilance, particularly when changes are made in production processes. There are several dangers: (1) the creation of unexpected effects; (2) the need to find them by testing; (3) the decision how long to go on testing.

Where bacteria are involved there is a greater risk, because they are anyway prone to mutation. It is also more difficult to test each one individually, but bacterial fermentation has been a feature of the food scene for a very long time. For science the challenge is to delve deeper rather than withdraw from the field.

So Sir Jonathon Porritt believes Lord Melchett was wrong to lead a Greenpeace expedition to destroy a trial with a genetically modified crop, on the grounds that it was anti-scientific. It prevented reasoned argument replacing a mutual slanging match. The jury's verdict in finding Melchett not guilty reflected public feeling more than an objective interpretation of the law. For although the government has made a serious attempt to approach the subject of GMOs with due care, it has failed to defend its position with conviction. The government's attitude is generally regarded as positive, but anxious to give no offence to environmentalists promoting the Precautionary Principle. Yet in other situations Tony Blair has been ready to do what he believes to be right at the expense of his popularity. Many of their critics would prefer to see GMOs condemned without trial, because those in the public eye are sponsored by multinational companies, and are condemned for their commercial ambitions. Yet such critics are prepared to accept the actual and potential benefits to medicine, but deride the way staple crops might be enhanced for consumers in developing countries. So it is high time to return to the objectivity of science in establishing the realities of the way nature works. The government is accused of bias, when it is in fact trying to establish where the weight of scientific opinion lies. Now that the report of the government trials has been issued, it is interesting to see those who trashed them quoting what they believe supports their viewpoint.

In issuing the report Professor Chris Pollock, the chairman of the steering committee overseeing the trials, declared they were 'the largest agro-ecological experiment in the world'. They had demonstrated clearly and unequivocally how tight the associations are between agriculture and the environment. So the trial results are likely to affect future policy decisions whether GMO crops are grown or not. 'Monitoring the effect of the Entry Level scheme (a widespread new scheme for environmental farming) can now be done on a scientific basis as these results are a benchmark.'

In the case of oilseed rape, the GM crop showed more grass weeds but fewer broad-leaved weeds than the conventional crop. The significance of this is that the broad-leaved weeds are more important to wildlife – birds, bees and butterflies. The fact that grass weeds and grass seeds were much more prevalent in the GM crop was due to the late application of glufosinate, giving a relatively poor control compared with the herbicides applied much earlier to the conventional crop. Clearly there is a trade-off to be made between crop yields and the reduction in biodiversity. But it needs to be emphasised that the differences found in the trials had nothing to do with the way the GM crops had been modified. They simply reflect the differences between the two herbicide regimes.

More will be said on the practicalities of these issues in the next chapter, but it is worth considering whether biotechnology and organic farming should really be classified as incompatible. Biotechnology has three main areas of activity:

(1) Tissue culture – the ability to regenerate plants from a single cell, which is illustrated by Dr Ian Robertson's work in Zimbabwe, cited in the next chapter.
(2) Molecular biology – the isolation and description of particular genes.
(3) Transformation technology or genetic modification (GM).

Mike Gale of the John Innes Institute has, for example, speculated on the needed replacement of the Cavendish banana. The most popular variety in public taste, it is apparently threatened by disease and degradation. But since bananas are normally multiplied by vegetative reproduction, GM may be the most practical route for producing a new variety. Some scientists also contrast the blanket hostility to GM crop varieties with the acceptance of those produced by chemical and radiation mutations. Nigel Halford of the Rothamsted research station believes that attitudes will change when new varieties are produced which offer clear benefits to the consumer. 'If such varieties are successful, some consumers will want them produced under organic conditions' (*Farmers Weekly*, October/November 2004).

Biotechnology is a huge and growing field, and the world of science is essentially an international one. It is already global, but it is not already tied solely to commercial interests as some would appear to believe. Although it has to be said that the UK government's cutback on financing research was probably a serious mistake, not so much on economic as political grounds. Government simply can't afford not to be involved in the way science shapes our lifestyle, because it concerns the public good. It concerns decisively the impact we make on our planet in the future, and is showing itself to be a major preoccupation of modern society.

So government should have the courage of its convictions in sponsoring scientific research, and in curbing those who consider themselves justified in going beyond the law. But even more important is the need for scientists themselves to marry the detail of their work with the wider vision of the Creation to which it contributes. Professor M.S. Swaminathan from India, the internationally known agricultural scientist, suggests that there is 'a growing mismatch between the rate of progress in science, particularly in the area of molecular biology and genetic engineering, and the public understanding of its short and long term implications'. He notes the urgent need for institutional structures which can inspire public confidence that the risks and benefits are being measured in an objective and transparent manner.

> Scientists and technologists have a particularly vital role to play in launching an ethical revolution. With a rapidly expanding Intellectual Property Rights (IPR) atmosphere in scientific laboratories, the products of scientific inventions may become increasingly exclusive in relation to their availability, with access being limited only to those who can afford to pay.

Professor Swaminathan proposes that the UN could explore the possibility of establishing an International Bank of Patents for Peace and Happiness.

> Scientists and technologists from all parts of the world could be encouraged to assign their patents to such a Bank, so that the fruits of scientific discoveries are available for the public good. Such a Bank would stimulate scientists to consider themselves as trustees of their intellectual property, sharing their inventions with the poor in whose lives they may make a significant difference for the better.

In the wording of such a proposal, I am struck by the choice of Peace and Happiness, which do not usually mingle with scientific jargon. It suggests an association with an international purpose which, even if unexpected, should be universally welcome. The late J.R.D. Tata once said he didn't want India to become an economic superpower, he just wanted it to be a happy country. That rings a bell with ordinary people, but quite how multinational companies should adapt remains to be seen. Suffice it to say that it would certainly qualify as globalisation with a human face. For whatever the shortcomings of the UN, it represents a world forum which is still in evolution, and which has agencies grappling with issues in education, health, trade etc. What is more, there is access also for non-governmental organisations to express their concerns, and no one power can dominate except by the negative use of veto. So while it may be too much to say that it commands universal respect, it certainly does have respect, and the opportunity to articulate an ethical position in current conflicts. Such a need is widely understood, and could be the seed of trust in what is to come.

To go a step further, it could also remove many of the fears in the public mind about modern science. The commercial pressure to keep companies in the forefront of their field has dictated both the need for a high investment in research, and a protected period in which to exploit the knowledge so garnered. If the knowledge is to be shared, that period of protection might be virtually eliminated. But the credit for the work would still be registered and it should not be too difficult to ensure that it earned a fair reward. Already traditional plant breeders receive royalties on the seed of varieties which they have developed, and credit is or should be given to the sources of plant material which may have been used in that development. But such recognition and practice needs to become universal. There would be clear advantage in an international agency where the process could be transparent to the public, instead of being the object of endless suspicion, which on occasion has been found to be justified.

It should also help to restore respect to scientists working in universities, and for that respect to be reflected in their salaries, which are so much lower than those in the commercial world. In the long run knowledge is beyond price, and there are arguable reasons for removing it from the market place. That would not be to eliminate competition, because the technology and know-how of its production would still be a major factor in the cost to the customer.

In an age of instant communication and the infinite capacity of the internet, it is something of an anachronism to direct the scale and scope with which knowledge is deployed into real products. It could be time for a swift transformation into a new way of doing business. It could even be termed a fresh approach to global development, where so far, despite the assertion that no one is in control, opportunities tend to be limited to those in the swim. The thought of science becoming in some sense a public trust would be attractive to many, but it is likely to be heavily resisted. Yet the point at issue is not control but liberation. Science is part of the open field of knowledge for all mankind, and enclosures should be limited to the specialist fields of technology and know-how. For science is concerned with the search for truth which may be incomplete, but cannot be falsified. It deals with a world where natural laws can be proven, but speculation on their significance may be very wide of the mark. Even commercial development may be corrupting, if it makes ethics a secondary consideration.

3

Biotechnology in agriculture

Roger Bacon is regarded by some as the father of British science, and it was he who declared that 'Nature is governed only by those who obey her'. It is a way of saying that we are not discovering and promulgating scientific laws, but simply learning and explaining the realities of nature around us. So the way ahead is to pursue developments consonant with natural law. But this does not mean that we must forswear any role in the direction of evolution as interference. In fact it could be argued that at this level a whole new task lies ahead of us.

For the development of agriculture and the domestication of plants and animals has, as Julian Huxley pointed out, put us to a certain degree in charge of evolution. Crops and livestock have been consciously selected for commercial characters, and their nature changed in the process. But it has been a slow progress, which has enabled it to shape social and cultural development to keep in step through the centuries.

Increasingly, however, modern developments have employed methods which have led to criticisms of manipulating nature rather than working with it. Sixty years or so ago, Sir John Hammond, the noted animal physiologist, had to be drafted in to speak to the bench of bishops at the House of Lords in justification of artificial insemination for cattle. A respected and persuasive figure, he was apparently successful in overcoming their doubts, and artificial insemination of cattle very quickly became an everyday part of the farming scene. It can hardly be called natural, but equally it cannot be called against nature when a study of the reproductive process has made it so highly effective. As a result only bulls of proven performance are used nowadays, which greatly accelerates the rate of improvement. Yet sixty years ago bulls were licensed after visual inspection, in an effort to prevent the use of poor bulls under the pressures to save money by false economy. So bureaucracy has been eliminated, and new knowledge has been put in charge.

Like many such steps in biotechnology, artificial insemination is a tool to be used, and it has brought clear benefits in the management of dairy cattle, and the pace at which breed improvements can progress. If there is a downside it can be seen in what might be called a bandwagon effect, when people flock to share in the benefits on offer without perhaps adequately thinking through what they want to aim at in their own particular situation. In the UK this is reflected in the domination of milk production by black and white cattle – first Friesian and now Holstein. Bulls are on offer with high and proven production figures which can lift a herd's average remarkably quickly. So it is understandable that they attract a wide following. Nor as yet can it be said that genetic choice is being seriously narrowed down.

In such circumstances it is easy to start a line of action, and become absorbed in it without much peripheral vision. It could be said that this has led to a lack of thought about the potential of other breeds, or the importance of other characteristics such as fertility, health and longevity. That can be the thin end of the wedge towards regarding cows as machines to produce milk rather than animals whose all-round performance and life must be taken into consideration. Yet ultimately such tendencies will become self-correcting, as those involved become aware of them. In discussing robotic milking, three times a day milking or overall herd performance at lower levels of intensity, producers become more aware of these other factors.

Such tendencies, moreover, are not a direct result of embarking on artificial insemination. A more recent tool in the shape of the computer could also be seen as a two-edged sword, in the sense that programmes can sometimes be adopted and become standard practice, without regular review of what they actually deliver. Thus complex mathematical formulae can be devised for animal breeding which give variable weightings to a range of characteristics designed to move in a desired direction. But those making use of such programmes need to remember that they are not cast in stone, and are open to adjustment – so they must be judged in the light of new experience and any changes in customer demand.

This may be particularly true in breeding sheep, where many breeds are available and desired characteristics are not necessarily highly heritable. I have myself been involved in these questions since I have always believed in recording performance, and have consequently had to look for others with whom to work. The advent of the Meat and Livestock Commission (MLC) provided the necessary expertise with computers, and programmes have been established which are now showing their worth. This has enabled group breeding programmes to chalk up clear improvements in performance, and even to permit comparison with other flocks previously considered to be ruled out by the inevitable differences in circumstance. The layman has to take a lot of this on trust, supplying the records, which tend to come back as gospel findings. But it is important to keep in touch with the underlying science through those willing and able to explain it.

With the costing of an enterprise there can be no argument, and it is extremely useful to have a benchmark of how others are performing. But in the quality of live animals something more than a paper appraisal is still necessary. Competition in show business has sometimes tended to devalue the importance of judging by appearance, but the eye still remains an aid in judging conformation and the living quality of an animal. Yet we have moved on from trying to create an outstanding individual, to raising the whole level of performance in a flock or a herd.

Dolly the sheep may have died at a comparatively young age, but the importance of her life was not as a farm animal, but as a demonstration of how a single cell is programmed by nature, and can be encouraged to develop into a complete and normal animal. It has led some of the scientists involved to turn to stem cell research in humans, and its possible implications for medicine. British researchers are now in the front rank in this field, and there is no difference in scientific terms between animals and humans. Medicine indeed is likely to be the chief beneficiary of much that we are learning through biotechnology, since farming is unable to carry the high costs which may be involved.

Also as Dr Harry Griffin (assistant director of the Roslin Institute, where Dolly was cloned) has pointed out,

> If you clone today's best bull, by the time the clone is mature, the original could have been mated with the best cows and at least half the offspring would be better. Conventional selection will grind out improvements year on year, whereas cloning simply stops progress and replicates the situation of a few years earlier.

Expense has to be proportional to the progress which will pay for it, and clearly investment in farming should always be looking at the long term, which probably means a modest return.

There is much misunderstanding over the word 'clone', and it is worth quoting from a 1997 essay by Dr Rosa Beddington (Mill Hill, National Institute for Medical Research):

> While cloning can offer the scientist important answers to fundamental questions about genes, it has a much older and very natural history which long precedes the sophistications of the modern laboratory. The word 'clone' comes from the Greek meaning twig, and there is a very good reason for this. For example, every chrysanthemum plant you buy at a Garden Centre is a clone of some distant and probably long dead chrysanthemum, which once supplied a side-shoot for rooting. Likewise, whenever you divide an overgrown shrub or successfully cultivate a houseplant cutting, you are cloning. In each case you are deliberately propagating a copy of the parent, and eventually over many years and hours in the greenhouse, producing a multitude of plants [clones] all genetically identical to the prized parent. Elm trees and other suckering plants clone themselves naturally, sending out subterranean roots from which new plants, of identical genetic constitution, will sprout. Deliberate cloning is as old as horticulture itself. Thousands of years before anyone understood the physical nature of heredity, specific genetic constitutions were preserved through cloning because they bestowed on the plant desirable qualities such as disease resistance, high yield and predictable growth. Cloning is as important to the production of fine wine, the supply of rubber and the fruit harvest as it is to the variety of an English country garden.
>
> Furthermore, natural cloning is not confined to plants: microbes and some insects frequently propagate themselves by producing genetically identical offspring without recourse to sex. The toothless mammal, the armadillo, gives birth not to identical twins but to genetically identical octuplets: every litter a batch of eight clones. There is nothing a priori unnatural about cloning.

What farmers are seeking in their livestock is robust health and predictable performance. They have advanced mainly by trial and error, the latter being attributable to excessive inbreeding to close relatives. But as Rosa Beddington points out, 'Disadvantages aside, the huge array of dog breeds illustrates that striving for genetic similarity and stability, contrary to popular belief, does not necessarily decrease diversity but actually often generates greater variety'.

Meanwhile in the sphere of crop production, plant breeding is beginning to combine genetic selection with genetic modification. But, as we have seen, much of the controversy which has been generated by this development has centred on the activities of particular companies like Monsanto, rather than the science as such. It has sidetracked the debate into being for or against GMOs, rather than what biotechnology can hope to achieve, which may cover both smoothing the path of the farmer and offering enhanced nutritional value to the consumer. For the layman, whatever is done must be seen and be understood in all its ramifications, so the trashing of trials is a particularly mindless form of protest. It could only make sense if you either think you already know the answer or simply don't want to know.

As Richard Dawkins, professor of the Public Understanding of Science at Oxford University, points out, the possible risks from GM crops tend to divert attention from definite dangers such as the misuse of antibiotics, which are already well understood but often ignored. He argues that almost every morsel of our food is genetically modified, though admittedly by artificial selection and natural rather than artificial mutation. A wheat grain is a genetically modified grass seed, and each varied breed of dog a genetically modified wolf. Those who deride the development of new types of staple crops with enhanced characteristics need to consider whether they are not in fact standing in the way of evolution, because the new factor on the scene is the evolution of mankind, which makes possible a consciously directed and purposeful course.

Evolution has been a long process of trial and error occurring with an apparently amazing profligacy and wastefulness. Dead ends have been commoner than breakthroughs, and Richard Dawkins comments that 99 per cent of all species that have ever lived are now extinct. Yet he is on more dubious ground when he states, 'The essence of life is statistical improbability on a colossal scale'. If the human brain is unique in evolutionary history for its capacity to conceive of a future, all that may be about to change; although this is not an invitation to abandon nature on the grounds of its short-term approach to survival. That would seem more than a little churlish, when human beings have somehow become an end product of evolution. It just puts a question mark over the assumption that nature's record has not been a progressive one, whether or not one postulates the existence of God. Louis Pasteur used to say, 'In the field of investigation, chance favours the prepared mind'. In everyday language it is perhaps tantamount to saying that just sometimes we make our own luck!

Nowadays also there is a distinction to be drawn between what happens in the laboratory and what happens in nature. The risks apparent in feeding poisonous substances in the laboratory may not always be reproduced outside, if other factors are also at work. Conrad Lichtenstein points this up in the story of the Monarch butterfly being killed by Bt GM maize, modified to poison the corn borer, a major insect pest. He wrote,

> The plants express a Bt protein– a natural insecticide of bacterial origin which some organic farmers spray on to their crops as the whole bacterium. So it is hardly surprising that butterflies are killed when fed such GM maize pollen in the laboratory. But in the field not having to spray chemical insecticides has actually led to an increase in Monarch butterflies.

Professor M.S. Swaminathan in India has suggested the need for a marriage between the scientist and the farmer in the field:

> An intelligent integration of Molecular and Mendelian breeding techniques will help to enhance the nutritive value of staples. By integrating pre-breeding in laboratories with participatory breeding in farmers' fields, it will be possible to breed location specific varieties and maintain diversity.

This would seem both to meet genuine concerns over loss of biodiversity, and the popular protests occasioned by reaction to the role of multinationals rather than the consideration of potential benefits for the malnourished.

Professor Swaminathan himself is a strong advocate of maintaining direct links between scientist and farmer. Even when responsible administrative positions made this difficult, his conviction never wavered. So it is interesting to find that, in Cameroon, Emmanuel Kamgono has produced a variety of maize – K525 – which has at least the same potential as varieties produced by the formal research system. A taxi driver who only began farming when he inherited land in 1993, he was taught the techniques of plant breeding by Alfred Müller, a German agronomist. It says much for his commitment and patience that he could produce a higher-yielding variety which also satisfied the popular taste – a factor neglected in many of the new hybrids.

As mentioned in the last chapter, the Farmscale Evaluation trials of GM crops have been an important beginning. So far no similar measurements have been made on the effects of organic farming, though even so-called superweeds would meet their death under the cultivator.

Since the current crops under examination are aimed at reducing herbicide and pesticide use, the trial results reflect the attempt to accommodate wildlife in conventional farming practice. To begin with the elementary, a farmer's aim is to free his crop from the competition of weeds or the depredations of pests. This will give his crop the chance to produce to its potential. So every crop is to some extent an interference with wildlife. One current trend is to encourage biodiversity on the headlands (field margins) to compensate for greater crop purity on the cultivated land, although this is not intended to detract from the objective of using fewer chemicals.

A first conclusion is that the contentious area remains the spread of GM plants into the environment. This is of real concern to organic farmers, and it is essential that their existence should not be threatened by GM practice. In a crowded country like the UK, with a high population in relation to the area of land, this looks like being difficult. Although the limits which will hold seed contamination below 1 per cent do not seem too great, there are, in the case of oilseed rape, wild plants which can keep the contamination alive. It is ironic that a tested molecular method is available to solve the problem, which has been dubbed the terminator gene. Its use has been ruled out because it would prevent farmers saving their own seed in the traditional manner, though they accept that this is not possible if they choose to use hybrids. Such negative publicity has made scientists afraid to use it to achieve isolation, but its degree of efficiency is also in question. This means that if absolute purity is to be an acceptable aim, it would seem to limit GM crops to the heartland of large acreage farms. With maize, on the other hand, no such problems arise, because pollination is restricted and there are no wild hosts. This underlines the comment of Sir Ben Gill, a recent NFU president, that each case must be examined on its merits, and that wholesale approval or disapproval on principle seems unreasonable.

The bigger question may be how to focus on initiatives which will be helpful where the need is greatest. That is to say the developing countries with growing populations, rather than the developed rich, who are or should be sated. Already good work has been done on sweet potatoes and cassava in supplying disease-free stocks through tissue culture, but much more could be done if it was seen as a priority. African farmers need peaceful conditions in which to work, but they also need the simple improvements which can help them to produce their way out of poverty. That should be the true goal of Land Settlement schemes, which must not only provide land for new farmers, but point the way into profitable production.

Dr Ian Robertson in Zimbabwe provides a good example of the kind of work I have in mind. He has launched a viable scheme for small farmers, many of whom are struggling to make a living. What follows is a brief summary in his own words, and although at present only on a small scale, it could be multiplied many times. In fact EU development funds in Africa could not be put to a better purpose.

> The average British 14-year-old schoolboy gets a pound a day in his pocket for pocket money. The average Zimbabwean worker (75 per cent) gets half that to feed and look after his family. It doesn't seem fair. Do life's chances depend on where you were born? Should it? Can we do anything?
>
> I am a good scientist earning a living teaching agriculture at the University of Zimbabwe. This year our farmers have planted about one third of the usual lands. Our country earns cash with tobacco, cotton and fruits and flowers; our people eat mostly sadza made from maize flour. The majority are eating only one meal a day already. The World food Programme says that by the end of the year 5 million will need charity to prevent starvation.
>
> Tapiwa Ruhode's research has shown that 3000 'born again' sweet potato plants planted in a 30m x 30m plot can feed a family of seven for 365 days a year. Now the Swedish Centre for Co-operation has stepped in to

pay Agri-Biotech, our company, to supply 20 nursery farmers in each of eight Districts with those 3000 starter plants. They grow them in August with irrigation for four months of sunshine and can then sell runners to over 100 neighbours in time to plant for a full growing season when the seasonal rains arrive in December. The cash earned is enough to buy a milk cow.

In the lab we dissect out the 0.25mm meristem (the tip of the bud – which is free from virus and other micro-organisms) and it takes nine months to grow it up in the test-tube into a regular 'virus-free' (born again) plant. From here we take it to the greenhouse plastic tunnels and take cuttings, lots of cuttings. These donors buy for 5 US cents. So we need good lab work plus good greenhouse management to deliver to good farmers.

The nurserymen then lift the tubers and sell them early when prices are good while the neighbours are growing for 'stomach-fill' for their families. Boy Ncube over two years has turned his US$150 delivery into US$5600 (£3000). He bought a milk cow from selling vines, and plans to build a brick house for his wife and buy a 'Bakkie' to carry his tubers to market. Over two years his 30m square grew to 3 hectares by replanting more runners. Our donors have estimated that for every kroner they put into the project our farmers – the 320 nursery farmers (20 x Districts x 2 years) of them – have earned four kroner – and fed their own families with quality food – high protein, decent amino acids and plenty of carbohydrate.

In the current emergency the Swedish Centre has asked Agri-Biotech to deliver 1000 plants each to another 1000 'beneficiaries' disadvantaged orphans, old folks who have lost their middle generation to HIV, or single parents. Our company employs eight BSc Agric graduates and ten greenhouse field workers and one secretary. The donors spent about US$300,000 over two years and our farmers cash in US$1,200,000. Agri-Biotech earned

US$50,000. We have only covered eight of 56 Districts so any one who wants to participate can contact us on agbio@mweb.co.zw

(July 2004)

Meanwhile in the developed world, though farmers share a number of difficulties, it is the other side of the coin. Already overproduction of food has forced consideration of alternative energy and industrial crops. When it comes to growing rapeseed oil for tractors or oats for working horses, who is to say whether one is more natural than the other? The diesel engine was specifically designed to run on vegetable rather than mineral oil, showing an early commitment to renewable energy. In a similar context, the current predicament of the sugar industry in Europe reflects the need to adapt to changing times. There may be opportunities for ethanol production to improve the environmental performance of petrol. Brazil has subsidised the production of ethanol, and made it an important feature of their enormous sugar industry (see Chapter 12). But biomass from willow or grass may prove a cheaper source of energy here. In any case the biggest contribution of biotechnology to such non-food crops is likely to be in medicine and industrial chemicals.

The advent of plastic and its deployment in an increasing range of industries has prompted many to brand it the most artificial and rubbish-producing expression of modern technology. Here in Herefordshire, polytunnels, which do so much for fruit and vegetable production, are a subject of growing controversy on aesthetic grounds. Although I rather liked the letter to the *Hereford Times* by a local artist, who confessed he appreciated the wavy outline of polythene, but couldn't stand the harsh yellow of flowering oilseed rape. But plastic is a chameleon product which takes varying forms, and many of them pass completely unnoticed. We should not underestimate its versatility, and the contribution which agriculture can make to the virtue of biodegradability. For the realm of rubbish disposal is beginning to become a science in itself, stretching from recycling to composting and biodegradation. Disposal by dumping is rapidly becoming outdated, so that packaging and other materials may simply add to costs and inconvenience, if their end is not catered for. This makes it important when planning a new product to consider its packaging and final disposal, as well as its use by the customer. That is how the natural environment works.

In livestock husbandry the production of drugs from animals is not new – hundreds of thousands of pigs have been killed over the years to supply diabetics with insulin. But the ability to add new genes to an animal's normal repertoire or precisely modify one of its own genes is known as transgenesis. As Rosa Beddington points out,

> it offers a more imaginative and less destructive way of producing drugs from the farm. Transgenes have been introduced into the nucleus of cow, sheep and pig eggs, and become part of the resulting animal's genetic repertoire, indistinguishable as far as it is concerned, from its own genes. Such genes have been designed to cause secretion of human proteins into the milk, thereby turning the milking parlour into a drug production line. In this way animals have been generated whose milk contains human proteins involved in emphysema and blood clotting deficiencies, and it has been possible subsequently to purify quite large quantities of these therapeutic products from the milk. Such transgenic farm animals would allow patients to be treated with 'on tap' human products, and the transgenic animals themselves would enjoy exactly the same lifestyle as any other dairy animal.

Cloning might come into the picture here if it could permit a better and more controlled way of producing transgenic animals than is practised at present. But it is the transgenesis which has been researched for over twenty years which would give it a role.

Unfortunately for the UK, our government has been slow to appreciate these possibilities, and to put money into research rather than reduce it. Research is always worthwhile where there is a vision and an aim, but the current doctrine of leaving things to the market is based more on convenience than experience. Even Monsanto is reportedly regretting some things it has done in the heat of battle, and is not dogmatic on the rights and wrongs of patenting in the context of genetics. Science is essentially a shared activity, and could be leading us to a reappraisal of the way research is conducted, particularly if the general public is to be included into some appreciation of what is going on.

Suffice it to say that, despite the criticism heaped on modern industrial farming, there are many signs that technical development is not by any means necessarily against nature. The amount of new knowledge accumulated even in my lifetime is sufficient to fill a library, and now compact discs have come to our aid in handling the sheer volume. But more importantly the practical application of this knowledge has survived the first rush to keep up with the Joneses, and is meeting a more searching and critical appraisal. Perhaps too it is better understood that the diversity in nature can also be expressed in the variety of human civilisation.

If we are to live in a world where change is a constant, it does not have to mean constant dislocation. Well-managed change should be an organic process which involves all contributors. Biodiversity allows for a variety of approaches with varying degrees of success. Progress will lie with the examples which are worth following. They will be compelling, but there must be no compulsion. Farmers have always been happy with that, which is at best the scientific way, while equally important it is also the democratic way. But it is a path which also calls for economic change, something a good deal more radical than adjustments to the current way of doing business. Economics must not be allowed to dominate and deform our future. Economists can help to produce the fresh thinking needed. As Nelson Mandela put it, 'Feeding the world is a moral and not an economic problem'. So if ethical behaviour is the preferred route to sustainability, the economists must make it part of our thinking.

4

A shift in economic thinking

It might be surprising to talk of a crisis in capitalism when a lot more people are visibly getting richer and more comfortable, but the question marks over the future are also stacking up. Communism failed because it promised 'a new type of man', and was unable to produce him. Stalin and Mao are monuments to that failure. Capitalism has never made such a promise, yet a great many supposed that the end of communism represented a victory for capitalism. It might be better seen as a challenge to capitalism to change direction. For such a change is an historic challenge which has both an economic and a political dimension.

If science owes its power and influence to what it can deliver, economics should demonstrate the ability to maintain a sustainable society. Tom Friedman, foreign correspondent of the *New York Times*, poses the question whether capitalism qualifies as an ideology, and then suggests there is no visible alternative. 'I don't believe there is an ideology which can remove all the brutality and destructiveness of capitalism and still produce rising standards of living.' But the weakness of his book *The Lexus and the Olive Tree* is that it is all about the Lexus and not at all about the Olive Tree. The latter indeed seems to be equated only with past tradition, rather than the rising foundations of a moral and spiritual society which values the ascent of man within whatever culture.

Ideology has been defined as a set of ideas embracing the three elements of a philosophy, a passion and a plan. As will be expressed in later analysis, a plan does not necessarily imply an aim to dominate, but it does imply aiming for a universal set of values. That is something which can create unity in diversity, and as Sir Jonathan Sacks, the Chief Rabbi, has underlined in a recent book, respect for the dignity of difference. But if, as some suggest, the process of globalisation may be the hallmark of the twenty-first century as opposed to the conflicting ideologies of the twentieth, it is as well to remember that ideology has always been about articulating a world view, even if it was often imposed by military conquest. Moreover as J.M. Keynes, the well-known twentieth-century economist, warned in his day, a world view includes much which lies outside the realm and influence of economics. So it would be folly to suppose that globalisation is only about the economy, important though that may be.

Adair Turner in his book *Just Capital* also makes this point, and is at pains to explain the current situation in orthodox terms. But he also puts forward some of the ways in which capitalism can adapt to current developments. He urges us to be sceptical of the belief that any one model of capitalism is the best. He suggests that so far as the US and Europe are concerned, the similarities of economic development are more striking than the differences. Yet even within Europe the differences in approach between different countries allow considerable latitude in the choice of a way forward.

So it is important to examine what we expect the economy to deliver. Per capita growth is taken by Turner to be a superior indicator than absolute growth. It is closely linked to labour productivity. Here the strength of the US position is maintained by higher overall employment and longer hours of work. The latter particularly is also a feature in the UK, where it has to make up for the failure to accumulate capital. We are now consciously paying the price for years of under-investment, which is particularly noticeable in comparison with France and Germany. But catch-up with others is always possible, and the changes or political trade-offs made in the overall picture are governed by the objectives we wish to favour. Perhaps the most fundamental is the redistribution of income to lessen the gap between rich and poor, which has been steadily growing. For the UK in the years 1973–98 the increase in real disposable income was – High earners 74 per cent, Medium earners 41 per cent, Low earners 22 per cent.

It might be possible to argue that this reflects meritocracy, if it weren't for the fact that the market is essentially amoral. Merit implies excellence and worth, which is not measured purely in financial terms. And the assertion that the market is the most efficient way of creating wealth begs the question whether generating billionaires is all that is mathematically required to measure market efficiency. The point about excessive boardroom incomes is not whether they would finance a needed raise for the lower-paid, but whether such leaders are setting themselves sufficiently challenging targets in terms of national and international responsibility. Amoral earnings become immoral earnings if they are awarded in the face of failure to achieve targets.

When Lionel Jospin, former French prime minister, said 'yes' to the market economy, but 'no' to the market society, he was making an important point. The growing gap between rich and poor is decried on all sides, but there is very little recognition of the changed attitudes needed to put it right. Noreena Hertz, of the Cambridge School of Business Management, puts forward the thesis that instead of fighting over territory, nations now battle over access to markets and economic freedoms. She supposes that the great wars of the twenty-first century will be trade wars fought with the weapons of commerce. She reports that Echelon spy satellites are now being used extensively for commercial purposes, while Bill Clinton, when president, decreed that industrial espionage should be one of the main tasks of the CIA.

Protests against globalisation are implicitly against this sort of activity, and the anti-war demonstrations evoked over Iraq reflect the widespread lack of faith in trying to settle matters by force or by economic muscle. Globalisation should instead be opening the door to concepts of Fair Trade as a basis for Free Trade, and the meeting of need rather than the sometimes artificially stimulated demand, which often comes close to greed. As Noreena Hertz also points out, the ordinary consumer turns out to be interested in investing in the common good. People are prepared to boycott unethically produced goods such as clothing from sweat shops. They are also prepared to pay more for ethical products such as Café Direct, which now claims 18 per cent of the UK market. Such consumer power is market-based, so it is effective where consumers can persuade a company it is in their financial interest to recognise it.

The growth of such initiatives is undoubtedly significant, even if it doesn't so far amount to a major alternative business model. Yet as the Fair Trade Foundation is pointing out, it is counterproductive to buy goods at or below the cost of production of even the most competitive. Commodity prices must be raised if trade is to bring serious benefits to the producer. It elevates the profitability of farming to a key issue in making poverty history. But it goes further than that, because world trade figures show that the volume of intercontinental trade is tiny compared to that of intra-regional trade. Only 11 per cent of the EU's GDP is traded outside its borders, so geography is still the major factor in food trade patterns. It also represents the most logical use of energy in transport.

A SHIFT IN ECONOMIC THINKING

Global pressures on trade intensity have been driven by transport and communications, and they may have already run their course to the edge of new limits. In agriculture the concept of 'food miles' has already questioned the logic of sending food halfway round the world if it can be grown in the neighbourhood. Yet Richard Scase, a leading business consultant, recently stated, 'In food retailing it means the end of seasons and seasonal purchasing patterns... anything can be put on our shelves in 24–36 hours.' That is the world of luxury, where exotic food can be had in the Hilton hotel in Addis Ababa, even when starvation wracks Ethiopia. But more worrying is the declaration of Clare Short, former Minister of International Development, '... old ideas that focus on national self-sufficiency in food rather than national capacity to purchase the food that is needed, are deeply outdated' (September 2001). Such a judgement ignores the fact that, in much of the developing world, it is the farmers who will lead their countries out of poverty. Yet to do it, they will first have to feed themselves and their own countries before tasting any major benefit from exports. Moreover transport by sea rather than by air is the only economic and environmentally friendly route. Freezing and chilling makes a lot possible, but for many items it still cannot compare with fresh food. So a certain degree of self-sufficiency is still strategically relevant, if less needed by Japan and other developed economies.

This is something to remember in analysing the trends being established by the new high-tech economy. Turner notes that agriculture is in steep decline, and manufacturing is following it. But he points out that in the US, international communications technology is not expected to employ more than 5 per cent of the labour market. So in sheer numbers of people employed the new economy depends above all on increasing face-to-face services, such as health, education, leisure pursuits and home improvements. As Friedman emphasises, the key to the information revolution is not that it increases labour output per unit, but rather that it reduces labour input per unit. This suggests that, contrary to popular conceptions of the rat race, it should be encouraging shorter working hours and a less pressured lifestyle. Business consultant Richard Scase poses the question whether legislation to restrict working hours can be effective. Citing Scandinavia and France he concludes it can, though commenting that company psychology in the US and UK makes people want to work. Only when people are in their forties may they look for escape.

Two other growing categories of expenditure and employment are housing and transport. But the demand for productivity in professions like medicine and education raises questions which the market approach cannot answer. As Turner comments, 'The market economy is a tremendously powerful tool to achieve ends, but it should not pretend to reflect the full range of human motivation and aspirations'.

This is emphasised by Larry Summers, a former US Treasury Secretary and Nobel Prize-winning economist, who commented in 1998, 'The US has lately been extremely successful because the balance of economic advantage has shifted from command and control in favour of motivation and incentive'. But he went on to say,

> As a model for making the economic pie as large as possible, American capitalism is second to none. But as a means of creating a stable society and addressing the concerns of our citizens, the American model's superiority is far from clear... A child born in New York today is less likely to live to five than a child born in Shanghai. A young American black man is more likely to go to gaol than to college.

On a more personal note, George Soros, the financier, writes (*The Crisis of Global Capitalism*) that he could not have continued

A SHIFT IN ECONOMIC THINKING

to live as he did as a speculator, and been a complete human being. He has turned instead to setting up his Foundations to spend his profits creatively.

> I have lost the ability to operate within the confines of the market as I used to. I have dismantled the mechanism of pain and anxiety which used to guide me...
>
> Integrating the various facets of my personality has been a great source of satisfaction to me. I must confess however, that I could not have achieved it if I had remained an active participant in financial markets.
>
> As an anonymous participant in financial markets, I never had to weigh the social consequences of my actions. If I had had to deal with people instead of markets, I could not have avoided moral choices, and I could not have been so successful in making money.

His comments support the contention of David Korten (*When Corporations Rule the World*) that the global financial system has become a parasitic predator that lives off the flesh of its host – the production economy. Money managers stake their careers and reputations on making money grow at a rate greater than the prevailing rate of interest. But in the process money becomes separated from value. Investment is meant to be productive, but there is plenty of exploitative speculation. Kidder, Peabody and Co. announced that one of its senior traders had, over a period of two years, reported trades of over $1.7 trillion, with $349.7 million of profit. Yet nobody had noticed that only $79 billion of these trades had ever actually been made, and that they had resulted in $85.4 million losses. Accepting the trader's report, management had given him $11 million in bonuses, a promotion and a Chairman's award. They also reported the false profits as real profits to General Electric, the firm's parent company. It took more than two years to pick up the discrepancies.

Even apart from such major cases of fraud, corporate cannibalism means that increasing numbers make their fortunes purely in finance. Money can be generated first in a wave of mergers, and then a second time in the later spin-off of the giant into its component parts. Yet nearly all employment growth in the next ten years is expected to come from medium and small companies. A proportion of these will fail, but so far new start-ups have more than balanced the failures. So while capital can be transferred across the world at the touch of a button, it is far from clear that is necessarily for sound reasons. It is an area where the herd instinct should be resisted in favour of more long-term considerations, because in the short term there are only two ways of making money – creating debt or bidding up asset values.

For the general public neither of these is acceptable if it is detached from the task of building a viable business which contributes to the real economy. A revolution in the financial world may not be possible in terms of an instant transformation, but in terms of thought and purpose it is essential. Too much economic thinking has been adaptive rather than creative, and the eccentric examples of people like Ricardo Semler in Brazil need to be studied and understood, both for the fact that they work, and that they represent a new approach.

Long-term considerations will involve a readiness to modify old structures or to create new ones, but above all will nurture new thinking. It is a serious anomaly that with all modern technology can do to make life easier, we should still be driven to work long hours and to feel that we haven't time for the human things of life. It may not be the horror conjured up by George Orwell, but it is, in a subtle way, the dictatorship of the machine. It would never be part of an imaginary vision for the desirable evolution of mankind. Yet that evolution is what should be occupying our mind and thought.

It has perhaps not been considered the economist's job to shape the way things are done, and the essential needs of mankind tend to be listed as food, clothing and shelter. But if poverty is seriously to be a thing of the past, the means to develop and enjoy life must be included. Health and education are clearly in the forefront, but leisure and family activities must also figure. All these involve at root the fabric of civilisation, and the faith that progress in living is not just a matter of comfort and the freedom to indulge one's fancy. This means that economics can take us so far, but not the whole way. Yet income redistribution will prove to be an essential factor in creating a sustainable society.

5

Economic structures and the meeting of need

What would it take to produce an economic system aimed at meeting need rather than simply demand? As noted in the last chapter, need must cover the essentials of life plus something to invest in living. It is an aim which reaches beyond the world of brand promotion and advertising. Such issues are constantly being raised by those who protest against globalisation, but they may be mistaken in believing that they are powerless, unless they can make their point on the streets. It has been suggested that it is the inequality of power which is at the root of much discontent and protest today. But there is enough alternative thinking about ends and means to create a strategy for change. For right from the days of Adam Smith it was clear that self-interest alone could not sustain a progress in prosperity. For Smith himself was even more scathing about the selfish motivation of the rich than Karl Marx. So capitalist values have always recognised a system of ethics, which involves a good deal more than sanctifying profit or admiring the cupidity of accumulation.

Amartya Sen, the Nobel Prize-winning economist, suggests:

> The big challenges which capitalism now faces in the contemporary world include issues of inequality

(especially that of grinding poverty in a world of unprecedented prosperity) and of 'public goods' (That is to say goods which people share together such as the environment, climate etc.). The solution to these problems will almost certainly call for institutions which take us beyond the capitalist market economy. But the reach of the capitalist market economy itself is, in many ways, extendable by an appropriate development of ethics sensitive to these concerns. The compatibility of the market mechanism with a wide range of values is an important question, and it has to be explored along with the possible extension of institutional arrangements.

One major illustration of such necessary adaptability is the Indian conglomerate of Tata, which, whatever the ups and downs of the last hundred years, is still true to its founder's vision. Jamshetji Tata, after whom the city of Jamshedpur is named, always made the health and welfare of his employees a priority. He didn't live to see the city, but he left clear instructions to his sons on how it should be planned.

> Be sure to lay wide streets planted with shady trees – every other of a quick growing variety. Be sure there is plenty of space for lawns and gardens. Reserve large areas for football, hockey and parks, and finally earmark areas for Hindu temples, Mohammedan mosques and Christian churches.

He also declared: 'In a free enterprise the community is not just another stakeholder in business, but is in fact the very purpose of its existence'. One hundred years later, Tata was the first company in the world to implement a sustainable human development index submitted in June 2003 to the United Nations Development Programme.

For over seventy years Tata Steel has been free of strikes, but Tata Engineering (Telco) suffered a crippling 48-day strike in 1969, which was followed by a series of shop-floor battles between

rival groups of workers. The two rival leaders, V.N. Prasad and
V.P. Singh, became reconciled during a conference at Asia Plateau,
the Moral Re-Armament (now Initiatives of Change) centre in
Maharashtra. A brief account of what followed is taken from
Michael Smith's booklet *Beyond the Bottom Line* (produced by the
Industrial Pioneer in association with *For a Change* magazine).

For Kiran Gandhi there was still the question of how to sustain
and multiply personal changes of attitude within the company.
When a vacancy fell open for a training officer in the Management
Training Centre, Gandhi took it, encouraged by Telco's
then head of management development, Nazimuddin Ahmed,
who had also been to Asia Plateau.

The two men began thinking about the need for 'self-motivation,' recalled Ahmed.

> Really motivated people are inner directed. They have a
> characteristic of their own and are not dependent on
> others. We decided to use the methodology we had picked
> up from MRA – the whole issue of inner change. We
> asked workers and union officers who had been to Asia
> Plateau to conduct a programme for a cross-section of
> people: divisional heads, supervisors, workers and opinion
> leaders. Kiran coined the phrase 'human relations at
> work' (HRW). It came just at the right moment, when
> people realised we had to change our ways.

The three-day HRW training courses, launched in 1982,
empowered the workforce to take initiatives, and led to the birth
of shop-floor quality circles, known in the company as 'small
group activity' (SGA). Groups of workers meet for an hour each
week to iron out production problems, as well as to discuss such
issues as how to tackle alcoholism, family debt and communal
tension in their townships or 'colonies'. By the early 1990s, 1300
such voluntary groups, each comprising eight to twelve people,
involved almost the entire workforce of 20,000 employees. Sarosh

Ghandy said that the SGAs had contributed substantially to productivity gains, saving Telco Rs 80 million (£1,900,000) a year.

Another key player in the birth of HRW was P.N. Pandey, one of four senior executives to visit Asia Plateau in 1978. He had taken a Rs 10,000 loan from the company on the pretext of buying a car – but had no intention of getting one, though he produced false documents to show the company that he had. Laying his career on the line, he felt he needed to be honest with the accounts department. Much to his relief he was not reprimanded, though he still had to repay the loan in 24 instalments.

Pandey, who had a reputation for being a tough and temperamental manager, also apologised to a supervisor, Jaswant Singh, whom he had insulted. They had not been on speaking terms for 12 years. Singh was 'completely overwhelmed', according to Pandey. 'He grabbed me in a hug and tears poured down his face.' A senior manager who witnessed the scene commented that Pandey had completely changed.

'This whole experience was formative in drawing up the content of the HRW programme,' Pandey continued. 'It was a live example of what human relations means.'

Shop-floor workers who have been through HRW are encouraged to be on its training faculty. At one afternoon session, a chargehand from the toolroom, R.B. Singh, conducted a programme for 20 production workers. On the agenda were alcoholism, absenteeism, punctuality, synchronising work functions, responsibility versus blame, and even blood donation schemes. Other HRW inputs include a session on conflict resolution, a discussion on the life and values of Jamshetji Tata, and an industrial film from Brazil made by MRA (Moral Re-Armament).

Sarosh Ghandy said:

> The main reason why we launched into our training using the approach of MRA in such a big way was to try and improve our managers and workers as people.

> We have also had two or three dozen swamis (Hindu priests) giving talks on various aspects of human behaviour.

Pandey added, 'HRW has led to a total attitudinal change among the employees at all levels'. In the four years that he was Telco's head of industrial relations, disciplinary actions against employees fell from 260 a year to 60. Man-hours lost in stoppages declined from 10,000 to none, and absenteeism, once chronic, was virtually eliminated.

It would be fair to infer from this that a new culture in capitalism can do a great deal to remedy current deficiencies, as well as defining the need for something more. Such a conviction may meet its most practical test in the further development of the World Trade Organisation, and the ramifications of relationships between the rich and poor countries. Concepts of fair trade and the essential standards of health and environment cannot be any longer excluded, particularly in the realm of farming and food production. It is perhaps in recognition of this that the Fair Trade Foundation has turned its attention to the application of its principles in European as well as developing countries' markets. At a time when western farm subsidies are under growing pressure for reduction or even elimination, it is worth examining why they ever came into being.

Farmers would certainly much rather receive their income from the market, and in the UK of the 1930s, government attempted to achieve this through Marketing Boards. In this respect Canadian experience seems to have been more durable, but post-war Europe was deeply concerned with raising production and establishing genuine food security. To raise production and guarantee a market proved much too contradictory to manage successfully. Technology advanced by leaps and bounds, while the numbers to be fed increased enormously but not within Europe. For a while the United States was able to follow its ambition to feed the world, and to show great generosity to those in need.

Yet it should have been more quickly accepted that those days were numbered. For too long complaints were being aired about the accumulation of surplus in butter mountains and wine lakes, and the pressure to reduce these led to dumping. It should have been recognised that, while real progress had been made in Europe, a new world view needed to take shape. This involves the understanding that each country needs to be able to develop its agriculture to its true potential, and that a concept of fair trade needs to be joined to the goal of free trade. Raw materials and commodities have always been at the bottom of the commercial pile. When prices are driven down, the buck stops there and only a cartel like Opec has been capable of proving otherwise. Fair trade establishes the principle that superior bargaining power should give way to the need to pay a price which leaves some margin over the cost of production.

The danger is that the farming industry may become stratified. One UK farm consultant suggests that as farm incomes remain under pressure, farmers will polarise into three main types.

> There will be the larger scale full-time producers who continue to grow in size, medium and small scale producers who add value to their produce, and medium to small scale part-time producers who increasingly rely on non farming income.

He comments that part-time farming is helping us to maintain a diversified farming structure. It allows many people to live and work on farms who would otherwise be forced to quit the industry. In fact rather more than half the registered farmers in the UK are already part-time in the sense of having another source of income, but it is not ideal.

Amid all the ingenuity required to make this work, the development of regional and niche markets has played an important part. It is also perhaps the most effective way to cater for the gourmet or luxury market. Although in addition to tapping into local patriotism, there is the important point that food miles are enormously reduced, with all the environmental benefit of reduced energy use.

Market research has long established how to measure a potential market both for its size and the sums of money it will be able to generate for particular goods. It remains an essential preparation before embarking on production. But the salesmanship currently employed to exploit markets becomes either more intrusive through the telephone or more appealing to greed through the offer of free gifts or cash rewards. In fact the pressure to advertise, and the money it generates, has even driven newspapers to produce weekend magazine supplements. These are given some content by the efforts of journalistic staff, but too often they are of dubious value. Sex and humour are exploited for imaginative titillation, but much advertising lost any connection with the positive qualities of the product a long time ago. So it becomes amusing to find Ronseal plugging the fact that 'it does exactly what it says on the tin'.

All this suggests a departure from first principles which mirrors in the market place the degeneration of financial practice, and the more dubious ramifications of so-called creative accounting. So it is time to take business ethics seriously and look at social responsibility rather than personal wealth. Interestingly enough, many of those who have accumulated great personal wealth are already doing this in seeking how to give it away. They have set up Trusts and Foundations endowed with their own capital to pursue charitable and creative aims. But the real point at this juncture may be how to make such initiatives a part of the mainstream rather than an individual exercise.

Even so change may come through evolution rather than by attempting a blueprint. Many individual businesses can point to experiences which are the forerunners of such change. Mr Charles

ECONOMIC STRUCTURES AND THE MEETING OF NEED

Denny, a businessman from Minneapolis, is one of the pioneers of business ethics enunciated in the Minnesota Principles. Describing his reactions to the fall of Enron, he tells how he belatedly became aware that the conflict between auditing and consulting existed in his own company. He had questioned it, and realised it required action, but had not pushed his questions to a conclusion. A niggle had remained on his conscience, but was drowned out by the pressures of the daily round and the desire for relaxation rather than further wrestling with the issues. His honesty on the subject will guarantee he is not the only businessman who has understood the dangers of such compromise, and will back his words with action.

That is perhaps the point at which a wider change can begin. Tom Friedman seems to be getting into deep water when he asserts – 'We need to demonstrate to the Herd (the providers of capital) that being green, being global and being greedy can go hand in hand'. The modern desire to proclaim a 'win-win' outcome to conflict is in danger of ignoring the need in many cases for honesty and recognition of wrong to open the door to change and a willingness to sacrifice a cherished point of view. A suitable outcome must always have positive elements for both parties in a dispute, but often the sacrifices which make them possible are just as important as the gains.

The imperative to stay ahead of the competition militates against this, because concessions are seen as a weakness. But hopefully the current trend towards alliances and partnerships will lead to a fresh evaluation of teamwork. Certainly the birth of a new understanding is the best guarantee against a return to the past, and represents a radical step on the democratic path. That is certainly true of the changes needed in the economic sphere, where trust has always played an important role, and one which has been undermined by many modern developments, not least the tendency to resort to the law.

Trust is built between people who know and respect each other. So it should be clear that it cannot be created electronically, and the present global financial market system is seriously flawed. It certainly seems to be true that international currency markets, which trade some $800 billion a day, do far in excess of what is needed to cover the daily turnover of goods and services. This money does not represent a real value, yet managers stake their reputation on making money grow faster than the prevailing rate of interest, though they are not personally liable if they fail.

The movement of companies to locations where wages are low is another example of an accepted economic imperative, which may on a longer-term analysis be less convincing in human terms. Korten cites the Stude Rite Corporation, a shoe company whose chief executive officer was sacked for trying to keep jobs in the US. Instead, production was moved to China, where wages were less than one tenth of those in the US. Sales have doubled and share prices have increased six-fold, being popular with social investors favourably impressed by the company's record of corporate giving. 'If the company had not done this it would almost certainly have been a target for take-over.' There are many examples nearer home, such as the Dyson vacuum cleaner company which moved to Malaysia, but if shares were not being publicly traded a different view might have been practicable. James Dyson himself, in his Richard Dimbleby lecture, spoke of China becoming the workshop of the world through its mantra of putting employment before profit. Partnerships and the long-term view are more important in manufacturing than in services, though such are the uncertainties invoked by rapid change that comparatively few are planning more than five years ahead.

If they were, there might be a question mark over moving industries to profit from cheap labour, when the whole aim in developing countries should be to raise wages. That is what will give life to new economies as increased purchasing power feeds off a rising turnover. Present practices are vulnerable to charges of economic colonialism, and hardly in tune with current opinions which emphasise partnerships and alliances. It is certainly not necessary to wait until the rich/poor gap is closed to establish a partnership of equals.

Ricardo Semler in Brazil has done just that, and proved that it is practical. Rather engagingly he says he knows it works in practice, but will it work in theory? Semco in 2002 consisted of 3000 people working in 3 countries in manufacturing, professional services and high-tech software. Wal-mart now uses Semco to count their inventory, manage their cooling towers, administer buildings and conduct environmental site investigation and rectification.

Its method is the ultimate in practical democracy, where everyone has a vote on how the company develops. Keep questioning and the pat answers will break down. But Semler makes clear that it is anything but an easy option, as the first requirement is to relinquish control. The boss has seen some of his pet projects voted down without redress. That is the measure of a revolutionary path.

The fact that it works is what has seized the public imagination, and led to it becoming a case study in seventy-six universities. Semler explains that

> Sustainability, productivity, profit, growth, new ventures, are all a by-product of worker balance. Everyone has a wealth of instincts, interests and skills which combine to form their talents. Some refer to it as a calling. We have been known to place 'ads' reading: 'We have no openings but apply anyway. Come and talk about what you might do for us, and how we might create a position for you'. The priority in hiring people is to seek those who will find a 'click' between their life purpose and the company's.

Though some might see it as a recipe for chaos, the company has been growing nearly 40 per cent per year without public investment. But Semler is keen to point out that if you grow too quickly, your culture may dilute to danger point. The rate of change needs to be managed. Keeping up with new developments is often the clinching argument for speedy action, but the short-term view will want to do things too quickly. 'Success begs for a definition to understand what we strive for. In nature growth is only seasonal, and perpetual growth is defined by cancers, which quickly become a danger.' Semler likes the question of one of his early professors, 'How far can you go into a forest?' Answer – 'Midway, then you start to come out'. At Semco, he says, 'We've concentrated instead on finding the right organic size for each of our markets. The aim has been neither continual growth, nor a takeover of the whole area, but creative adaptation in providing a service and filling a need.'

A further point about relinquishing control is that it means relinquishing exclusive rights to information. Trust and openness are the life blood of Semco – and everyone is invited to understand what's happening, even if no one may know exactly where it will lead. The trade union was asked to work out a simple set of figures on profit and loss for the easy understanding of the company workers, and there is no need for a human resources department, because it should be the concern of every manager. As Semler points out, trust and openness are vital to weathering Brazil's economic crisis, and its result is no lawsuits and no strikes. Only self-confidence makes freedom and dissent possible.

When flexitime was first introduced throughout the company the metal workers' union was among the greatest sceptics. As a result a special committee was created to iron out any problems which might arise. It never met. Instead easy control systems were founded on the philosophy that small groups of 6 to 10 people know all about each other, and can work things out. So Semco production units are never bigger than a few hundred people. As Ricardo Semler asks, 'Why do we think we are equipped to choose schools, doctors and mayors, but don't trust our capacity to lead ourselves at work?' And Semco is also famous for its stand against corruption. It refused a deal with a multinational which had countenanced a kickback to one of its client's directors. Yet as Semler comments – 'This multinational had a credo and mission statement that is beautiful to read, and would bring tears to the eyes of most.'

Another Semler question is, 'Why do we think that intuition is so valuable and unique, but find no place for it as an official business instrument?' The Planning Director of a major oil company told him that five years before he had predicted a price for Brent crude of $38.40 a barrel. The actual figure five years on was $18, and his intuition had put it at $23. The difference it made to the company's budget was $22 billion. Why hadn't he backed his hunch? 'Imagine me telling the Board that I had been sitting by the pool debating with my dog?' Why is he still in the job? 'I have the right to be wrong – but only so long as I am precisely wrong!' Intuition can't be bottled, so numbers reign supreme, and a man who has worked through the correct procedure cannot be held responsible for a bad result.

All of this can hardly be taken as a blueprint for changing institutions, but it raises questions which need answering and call for a much wider application. It is a shock to the system when a new executive hired at a good salary is briefed on the Semco ethos. He may find that he has no office, no secretary, no parking place, no official title and perhaps not even a desk. Yet the story is told of one such who made three or four major errors in backing ideas which failed, and cost the company immense sums. But the company stuck by him, and his next idea hit the jackpot. This was a story of human struggle and the effort to rise up from failure; perhaps the ultimate test of the democratic way.

Ricardo Semler himself is a charismatic figure, who became immensely popular in Brazil, drawing thousands to his public meetings. But he discontinued them completely when he realised he was in danger of becoming a cult figure, who would be called on to provide all sorts of answers for people unready to shoulder the responsibility themselves. But hopefully the revolution which he represents will continue to run its course. It is based on practical results in real life, and the interest which his example has aroused is evidence of its impact on business thinking. It is a way of putting convictions into practice, rather than following a conventional business model. According to some, IT has in effect wiped out middle management. Businesses are now finance plus ideas plus implementation. But however one may view things, it must be a positive encouragement when initiative and intuition can create new opportunities, instead of being strangled by fixed procedures.

The function of economics is to create a sustainable society answerable to human aspirations. Economists should be looking to liberate rather than put in a straitjacket. To show how a change of motive can enlarge the options. They should be closer to life than to law. Then they would have the potential to open a new international perspective.

6

Towards an international perspective

John Kay, in his classic book *The Truth about Markets*, suggests that 'History will judge whether Greenspan (US Federal Reserve Chairman) was the man who made millions of Americans rich, or the man who could not bear to tell them they had only imagined it'. Perhaps the underlying message is that he could have persuaded them either way. Kay's point is that the American Business Model (ABM) remains the working hypothesis despite some setbacks. 'Its appeal directly parallels that of the Marxist doctrines it has supplanted. Its prescriptions are not just right, they are inevitable.' Yet he contends that the ABM could never be a realistic model of how the American economy works, so it is misleading at best to urge it on others. 'Effective market economies are embedded in an elaborate social, political and cultural context, and could not function outside that context.'

In the coded language of international agencies, 'structural reform' means a move towards ABM principles. John Kay describes these as being of four kinds. First that self-interest rules, and self-regarding materialism gave us our economic lives. Second that markets operate freely, and attempts to regulate them by social or political action are almost always undesirable. Third that the economic role of government should not extend much beyond the enforcement of contracts and private property rights. Fourth that tax rates should be as low as possible, and the tax system should not be used to bring about redistribution of income and wealth. Kay finds this a naïve approach to human motivation, and comments that while greed is a human characteristic, it is not for most people a dominant one. He concludes that it is clear the American economy is so successful precisely because it does not conform to the Business Model.

As Kay sees it, the reality of the US economy is much closer to the embedded market, a term which he uses to describe the successful market systems of Western Europe. Such systems depend for their authority more on shared concepts than on law. Economic policy is as much about social attitudes and customary behaviour, as about law and regulation. So he sees the outstanding success of German vocational training, for example, as being down to the national institutions in which it is embedded. The authority of the state can be exercised to the extent that is seen as legitimate, but only to that extent. Yet, far from the theory of minimal government, productive economies have the largest, most powerful and most influential governments the world has ever seen.

All of this is a strong pointer away from fixed systems to the development of economic thinking as a means of contributing to a changing society, which is no longer driven by the idea that there is a virtue in the creation of wealth, no matter how it may be used. This might seem straightforward to the layman, but has come to be regarded almost as a sacred economic assumption. As a result it has become accepted in the public mind that the answer to poverty is any sort of wealth creation, regardless of what percentage may end up in the pockets of those in need. In fact the answer may be zero if it is hidden in a Swiss bank, but even going through the legitimate channels of aid, the leakage may be considerable to administration and planning as well as dishonesty.

Unless you visit village farmers in a subsistence economy, statistics can hardly convey the reality. It is not so much that the daily round may be dominated by tasks like fetching water or firewood, but rather that simple basic inputs can bring about so great a transformation. The NGO, Farm Africa, demonstrates this time and again with their distribution of dairy goats and chicken rearing kits (14 chicks, vaccinations and the hay boxes which will be their home). With the training which is provided, milk and eggs become available both for the family and to sell. In a surprisingly short time neighbours are benefiting from the offspring of the livestock, and daughters are going to school instead of having to stay home (www.farmafrica.org.uk).

At a slightly more sophisticated level one can trace the development of the production and export of dessert grapes in the state of Maharashtra in India over the last twenty five years. It began with a pioneer who studied the practices in California with the variety Thompson's Seedless. His successful pioneering quickly spread. One young man I met had been saved from debt and disaster by planting half an acre of grapes. The market expanded rapidly in the big cities of Mumbai, Delhi and Kolkata because there is a growing middle class in India whose lifestyle is on a par with Western affluence. So it was domestic demand which drove the development, even though it has also gone on to export. Today regular container-loads of grapes go from Mumbai to Felixstowe to supply British supermarkets, but it does not guarantee that the producer is reaping a commensurate profit for greater enterprise. It has been the result of a lengthy process, and such markets are by no means assured.

These experiences underline the fact that it is the farmers themselves who are a main road out of poverty, and should be recognised as such. So the basic foundation for such progress must include, where necessary, the protection of home agriculture and the provision of advisory services and training. It could mean that the exodus from the country to the towns can begin to bear some relationship to the number of jobs on offer. It should not escape notice that the outgoing prime minister of China (Zhu Rongji 2003) said that the number one priority facing his successor was to raise farmers' income.

Hopefully this may indicate that it is a mistake to assume that future economic patterns can be projected from past experience. This alone could have a marked influence on economic theory, which does not easily adjust to a world of multiple solutions. In the US, manufacturing's share of GDP is well below the European average. One estimate puts the GDP in the US as almost half devoted to transactional employment – that is to say, those not directly linked to production. It is a figure which has risen sharply, as it also has in New Zealand as a result of the sudden switch to free markets. But social pressures and social consensus are cheaper than law enforcement. That is how Japan and the Nordic countries run market economies with lower levels of transactional cost than the US.

Joseph Stiglitz, the American economist, argues that the IMF (International Monetary Fund) has deprived countries of choice on the way ahead. 'Globalisation often seems to replace the old dictatorship of national élites, with new dictatorships of international finance.' Without the cancellation of debt, many of the developing countries simply cannot grow, and he questions whether money paid to President Mobutu in the Congo can justly be demanded of his successors. It was paid to keep a corrupt leader aligned with the West, but it was known that a large part of it would be appropriated by him personally.

Stiglitz, however, considers that globalisation can be a force for good. The globalisation of ideas about democracy and society has changed the way people think, while global political movements have led to debt relief and the treaty to ban landmines. Those are an indication of the political will to achieve concrete change, and the free market ideology needs to be replaced by applying economics to produce the right answers. But each country must begin its own reform, as shifting the international balance will take time. Each must assert its right to make its own choice, particularly in relation to corporate misbehaviour and speculation. If speculators only made money out of each other, it would be an unattractive game. What makes speculation profitable is the money coming from governments often supported by the IMF. When the IMF and the Brazilian government spent $50 billion maintaining the exchange rate at an overvalued level in late 1998, where did the money go? Much of it went into the pockets of speculators, because, on balance, speculators make an amount equal to that which a government loses.

In standard market economics, if a lender makes a bad loan he bears the consequences. But repeatedly the IMF has provided funds for governments to bail out Western creditors, so that the creditors become less worried about the borrowers' ability to repay. Then the bail-out has amounted to free insurance. Though, of course, it is not officially admitted or countenanced, there has in effect been a change of mandate. The aim has changed from serving global economic interests to serving the interests of global finance. If a country defaults the IMF is the preferred creditor, and although capital market liberalisation may not have contributed to economic stability, it has opened up vast new markets for Wall Street.

It is also true that the US Treasury has tried to give the impression that the US taxpayer pays for IMF bail-outs, but the IMF nearly always gets repaid, and the money comes ultimately from workers and other taxpayers in the developing countries. Before September 11, 2001 the US Treasury even defended the secrecy of offshore banking centres. But secrecy undermines democracy, because there is neither participation nor accountability. Yet reforms in offshore banking only got moving after September 11, because the centres owed their existence to deliberate policies in the advanced industrial countries, pushed by financial markets and the wealthy.

On the more positive side, the EBA (Everything But Arms) Treaty of the EU is a step in the right direction. The EBA initiative removes all quotas and tariffs on goods from the group of 49 Least Developed Countries (LDCs), so granting free access to the European Market. Most of us would probably agree that economic justice requires the developed countries to open themselves up to fair trade, and equitable relationships with developing countries. This entails positive initiatives taken without recourse to the bargaining table, or attempts to extract concessions in return.

Although accusations of neo-colonialism through economic means are not always well founded, it is clear that large and rich countries do exert hegemony over others. Graham Turner in the *Daily Telegraph* (June 2004) observes that 'Nowhere in America is there a greater sense of the sheer intoxication of supreme global power than at Investment Banks'. A senior executive in Goldman Sachs told him: 'What we have in New York is unfettered capitalism, and that involves killing each other. A lot fall by the wayside. It is a Darwinian model and the survivors are very fit.' Yet another senior man in the same firm observed that 'at some point we have to move from success to significance'.

Apart from the fact that Darwin might pose the question, 'fit for what?' this attitude assumes that unfettered capitalists are no longer answerable to government. That is how some might like it to be, but the realities of political democracy dictate otherwise. Democracy contains all the weaknesses of humankind, but it also has the creative potential to develop fresh visions, and new concepts of society. If one thing is certain, there is a world beyond today's economic horizon and many people are longing to find it. In India and Africa, where perhaps the pressures of harsh reality more easily break through rigid economic theories, good governance is becoming a growing theme. The thought that capitalism has yet to find a civilised expression begins to take root, and the changes needed are fundamental. But they depend on a matching renewal in politics. Government has to develop the vision and make the pace for national and international change, which is based on the personal change of moral integrity.

7

The political dimension

The financial journalist William Keegan (*The Spectre of Capitalism*) has suggested that capitalism is not a system or an ideology but a way of life. He concludes this from the way capitalism works in different countries, and how each culture puts its own stamp on things. But his conclusion is that economics and politics have a symbiotic relationship. So that if people's lives are left to the vagaries of the free market, capitalism is indeed a dangerous spectre, for government needs to have a strong role if the relationship is not to become parasitic, or the economy to become the only focus of a country's life.

Democracy is essentially an evolutionary process, and respect for that process should guarantee that when mistakes are made they can be righted. The great French journalist Hubert Beuve-Méry, editor of *Le Monde*, in rebuking a cub reporter for inserting his own opinions into an article, spelt out the code he practised. He emphasised that the journalist's role is to provide his reader with all the necessary unbiased information which will enable him to make up his own mind about a given event. 'Only then,' he would say, 'will the reader become a citizen, and only then will we have a democracy'. That sets the criterion for a well-informed public.

THE POLITICAL DIMENSION

The fact that leaders and public men come and go as the reins of power change hands should serve to underline the fact that that no one is indispensable. But too often in recent years, those who have been in power for a long time have begun to suffer a certain 'folie de grandeur'. While those who have an agenda to complete may cease to heed the voice of the people or perhaps to identify it correctly. Mahatma Gandhi declined to play any role in government when India became independent, although he continued to wield immense influence until his death and beyond. In doing so he set an example of what has come to be called 'servant leadership', which does not depend on leading from the front, so much as the power of example and of practising what you preach. At a recent conference on good governance in Africa, held in Ghana, the Revd K.A. Dickson, president of the All Africa Conference of Churches, outlined this path when he said: 'There is a viable alternative to ruling, and that is serving. Exercising authority for good involves putting the people first, by engendering faith in their own ability to move forward in dignity'.

It is becoming increasingly clear that democracy at this level requires a high degree of honesty and transparency. But the process is not helped if government puts its own spin on the information provided, and its critics display their own bias in reply. It then becomes increasingly difficult to draw a line between cynicism and scepticism. Yet to do so is vital to the sustainability of democratic life. Whether or not it is hailed as a Third Way, it would seem that it may be high time for fresh political standards and priorities. There are some subjects which lend themselves to building a consensus, and could benefit from an all-party approach. Pensions and climate change are two of these, and need to be longer term than a single government. So it would help if the possible remedies were developed by collaboration rather than competition.

In this context it is interesting to consider the basic programme of the SPD (social democratic party) in Germany, which recognises the impact of 'post-materialism' in developed countries. It defines this as the level of prosperity at which people become more concerned with quality of life than with their financial state.

Quality of life brings us immediately to moral and spiritual considerations, and the very widespread desire to do something about the growing gap between rich and poor, which more than anything else disfigures our social progress. It needs to harden into a determination to close that gap. The response to the Jubilee 2000 campaign for debt relief illustrates what can begin to happen in response to popular demand, and a well-conceived campaign. Although it may still have a long way to run, tangible results can already be seen in the first countries to benefit from the cancellation of bilateral debt, such as the development of education in Uganda. Primary school places have increased from 3 million to 8 million in six years. While those who have had an education are shown to be less likely to get Aids, and more likely to get a job and raise a healthy family. The commitment secured from Gordon Brown and other political leaders ensures that the campaign will continue, and prospects for multilateral debt relief have already begun to be addressed at the recent G8 meeting in Gleneagles.

Anthony Giddens in his book *Summary of the Third Way* feels that the Welfare State should be replaced by the Welfare Society. This is more than a cosmetic use of words, because it implies democratic rather than State responsibility, which would be more inclusive and have less of a handout image. It also recognises the immense amount of voluntary activity which already takes place. In 1991 there were 160,000 charitable groups registered in the UK, with 20 per cent of the population engaged in some form of voluntary work, while in the US some

40 million Americans belong to at least one small group which meets regularly. Many of these are people with a common professional interest or similar concerns, who meet together to pursue 'a journey through life'. People want greater involvement and meaningful work, but they are not finding it in the political scene and the confrontational approach. When the Conservative Party seeks to project itself as a party caring for the downtrodden, it is clear that the old class divisions between right and left are no longer a defining measure of political philosophy, even if such divisions are not yet eliminated.

The Third Way does not seem to be intended as a blueprint so much as a fresh lens through which to look at policy. It is an attempt perhaps to transcend both old-style social democracy, and the traditional liberalism which has lost its shape by becoming too elastic. It could even be seen as an attempt to take up the battle to relate policy to the needs of an historic evolution in human nature, which is essential to opening the door on a new world. For while economic globalisation sometimes breaks down barriers by riding roughshod over them, politics have remained embedded in the nation state. But the growing and inescapable interdependence between nations raises the need for a new concept of national interest. The arguments over sovereignty, and its theoretical loss through the exercise of international voting, tend to be a brake on such thinking, whereas the sovereign nation might be better defined by the integrity of its own way of life, and the example it sets within its own borders. This would be the kind of leadership to end for ever the dreams of empire, and the dominance of a superpower.

It is the kind of thinking which has driven the development of the EU and the UN, which, however imperfectly, are seeking a moral and just framework for international relations. Tony Blair has observed that the growing global interdependence has made it more difficult to draw a line between home and foreign affairs.

Increasingly we have to consider the way society at home may need to take account of a global society: not a clash of civilisations but the need for a world civilisation, for want of a better term. Far from society being a shadowy concept, it is likely to become a dominant issue in moving beyond the thrall of economic imperatives to a more creative view of democratic freedom.

In this sense the political landscape is already changing, and the old concepts of left and right begin to become irrelevant. Tony Blair and Gordon Brown, whether or not establishing a Third Way, are attempting to go straight ahead rather than left or right. So although this means a redistribution of wealth and opportunity, it is undertaken not so much as a class doctrine as because it is morally right. This, of course, is also the basis on which a nation's record on human rights is being put under the microscope. Not all the judgements made will necessarily be fair, but it represents a new preoccupation with the life of the individual, not for their patriotic obligations but for a person's own destiny and fulfilment.

This is the crux of the matter in seeking a democratic future, where the individual is empowered to take part, but is also mindful of the greater good. 'As I am, so is my nation' may seem a simple theme, but it has far-reaching implications. It certainly leaves no room for political apathy. As Noreena Hertz (Cambridge Business School) points out,

> It has been said that until the State reclaims us, we will not reclaim the State. But this is hardly a democratic text. Protest as a catalyst of change seems to be increasingly effective, and ordinary people may even have influence as individuals, if they have commitment and stamina.

Yet this presupposes a cause bigger than anyone's personal preferences or preoccupations. Democracy asks a lot more of us than simply casting a vote. People need not have a party political commitment, but they must have a stake in the life of their country, and a purpose to pursue. To be limited to an ambition for personal success is the death knell of democracy, because it leaves out a vision for others, and becomes exploitative. Clarity on this point is the key to healthy politics, where hierarchies and cabals are squeezed out by new forms of transparency and teamwork.

This is not to decry the value of institutions built through long years of experience, but rather to affirm that they must also be capable of reinventing themselves. Ricardo Semler, the self-confessed maverick of Brazilian business, has amply demonstrated that doing the opposite of conventional wisdom can pay off. Those who persistently ask questions, and follow apparently outlandish paths, can equally come up with the goods. For him it is the ultimate in the practice of democracy, and he hopes to convey that work is not just a way of earning money to indulge our leisure, but an integral part of a full and rounded life. It is in fact an affirmation that work, however unpromising its character, can still be a vocation, and that is something which adds meaning to the democratic path.

It could be said that democracy demands a continuing evolution in human nature, and without it problems will multiply. So the role of the individual in playing an active and not just acquiescent part is dependent sooner or later on the voice of conscience over-riding human expediency. In the last century governments came to recognise the validity of conscientious objection to military service in time of war, and it has sometimes even been extended to examples of civil disobedience. So there is a recognition that people may be answerable to a higher authority than the State, and that, in the final analysis, moral and spiritual values are paramount. Although this can degenerate into an argument about who occupies the moral high ground, there is at least a recognition that such an argument exists! It carries an important message for our future development at a time when religious faith is commonly so undervalued, and has been replaced by widespread indifference.

It is a time to look afresh at some of the issues currently occupying the political scene, and the possibilities of a new motivation. If democracy is properly understood and practised, there is no reason why faith in politics should not be instantly restored, because it really doesn't depend simply on the politicians themselves. It is a concern and a challenge to everyone.

In some countries, grass-roots Clean Election Campaigns have paved the way for a focus on dealing with corruption and dishonesty. In Kenya the retirement of Daniel Arap Moi and the election of President Kibaki was one such example. It was a major political transition, because it also coincided with the drafting of a new constitution. In the lead-up to the election, campaigners urged communities to identify credible and honest leaders in their areas and encourage them to stand for parliament. They also spoke on hundreds of occasions on radio and TV talk shows, in church pulpits, in schools and at public gatherings. It set a positive and fresh tone which was widely noted, and in any democratic framework such action is possible, though it is, of course, only a beginning. There needs to be an ongoing strategy for change, which is the challenge examined in the next chapter.

8

The current political challenge

It would be fair to say that, despite the many authoritarian régimes still in existence, democracy is considered to be a worldwide goal and the only basis for sustainable free government. Democracies are nowhere perfect, but they open the door for change and evolution. They are also the essential guarantee of civil liberties.

Hitler sought to close down such options and ensure the Nazi vision prevailed. Stalin, likewise, was determined to see his view of communism followed throughout the Soviet empire. Such was the sequel to the First World War, a war to end wars, which quickly raised the spectre of failure to achieve any such thing. Rudyard Kipling, whose popularity endures against all expectation, had already written an epitaph for those times.

> The tumult and the shouting dies;
>   The Captains and the Kings depart:
> Still stands Thine ancient sacrifice,
>   An humble and a contrite heart.
> Lord God of Hosts, be with us yet,
>   Lest we forget – lest we forget.

But much had been forgotten even by the 1930s, and idealism chased a hundred shadows while tougher minds turned to the pursuit of power. Yet in that inter-war period was born in North America and Europe a new vision for the world, and the faith that ordinary people could make it a reality. In thousands of situations it took root as a commitment for life, which endured for all who survived the Second World War. Refined and matured by years of battle it focused on winning the peace, and brought to the task a body of people determined to see a better settlement than Versailles, and the League of Nations. For me personally it was embodied in the Oxford Group/Moral Re-Armament (now Initiatives of Change), which helped me face the inadequacy of my own way of life, and made me understand that personal change was only the beginning. It was not for my own comfort, but intended to be a foundation for social, economic and international change. That was the logic of the divine plan for those who committed themselves to find and follow God's leading. There was an urgency about the task which is still embedded in my heart, a kind of second nature, even though the climate of our times has changed. For while there has been a sea change in the timescale for building a new world, the need and the feasibility have remained articles of faith. Nor can they ever rest on someone else's commitment. To dedicate one's life is perhaps every bit as testing for the living as for the dead who gave their all.

How far modern history illustrates the answers beginning to take shape, and how far they are still to be articulated, is a matter for debate. It may even be questionable whether a single lifetime is long enough to measure an undeniable trend. But those who have lived through a slice of history are almost bound to think so. Public reactions to the war in Iraq have sprung from a variety of convictions, and would quite certainly have been more effective if they had reflected moral standards grounded and accepted in our national life. Yet they do reflect a clear rejection of military means as a sensible or successful way to combat terrorism or spread the democratic gospel. In Afghanistan terrorists had established a clear base for their operations, but in Iraq the position was different. Any valid change depended on diplomacy, and the mobilisation of international opinion. To take unilateral action meant sacrificing moral authority. There can never be sanction for an international police action without consent, even if Western thinking may support it. We have to face the current failures of diplomacy, and measure the action needed to redeem them.

It is easy to underestimate the profound changes which took place in the years following World War II. The British diplomat A.R.K. Mackenzie was then in the early years of his career. He comments on the San Francisco Conference which set up the UN in 1945 as one of the most exciting events of his career.

> There was a remarkable spirit in the air. World War II was coming to an end. Victory in Europe was celebrated three weeks after we arrived in San Francisco. The delegates, many of whom had assembled from war torn countries with blackouts and food rationing, found themselves enjoying not only the Californian sunshine, but the most enthusiastic of welcomes. San Francisco treated them almost like film stars. The result was that we worked in an atmosphere that was almost euphoric, and this idealism unquestionably helped us in our work. It was little short of a miracle to agree on the complex text of the UN Charter in two and a half months. Today it would take years.

Such was the prelude to Franco-German reconciliation, and the transformation in Japan's outlook on the world. At the same time the spread of independence and freedom to a growing number of colonial lands added a stream of new members to the UN. Not all were prepared for the opportunities offered, but amid the pluses and minuses of the 1960s, important commitments were made. Yet something was lacking, which sapped the urge to go further. Youth was celebrated, but not in terms of the self-discipline needed to bring a new society to birth. The permissive society, whatever the intentions of those who coined the phrase, singularly failed to measure up to world needs. Human rights had yet to be balanced by acceptance of responsibilities. A plethora of potential alternatives produced no commanding or overarching idea. So few were ready for the changes which came with the fall of the Berlin Wall, and for the statesmanship needed today.

Yet the feeling remains that there is a destiny for mankind which is worth finding, embracing people of all creeds, races and cultures. A moral and spiritual vision is the one requisite for a positive outcome, and it may depend on the kind of welcome we give to the initiatives offered. A consensus on such a vision might well be possible across a wide international spectrum, but it would have to be conceived outside the realm of power politics, as it is popularly conceived today.

The US at the height of its power has accepted to take a lead in the world, and is widely looked to for support whether military or economic. In the aftermath of the Second World War, there was much to be grateful for. The Marshall Plan became a byword for putting nations on their feet economically, while the man who gave it his name was known for his integrity and his modesty. So how does it come about that so much gratitude has dissipated, and anti-Americanism has been too easily fuelled by bitter criticism and downright hatred?

Perhaps the heart of the matter was to allow the struggle with communism to become only the struggle with an evil empire: a competition in military build-up. Ideology was examined, but the US was too convinced of the moral superiority of the American way of life to believe that any great change was needed or even desirable. Much has been written about Vietnam with hindsight, but perhaps the biggest lesson should be seen as the failure to listen to the Vietnamese themselves, and to read men's motives. The decision to unleash the Vietnamese rebel generals and destroy the Diem government came from the State Department and was not a united American decision. It amounted to sanctioning violence for political ends, betraying an ally, and in the event made escalation of the war inevitable. It was the point at which the Vietcong strategy of spreading discontent in the south gained credibility. It was also the beginning of the establishment of anti-American feeling in millions of hearts. After 9/11 Cardinal Egan, Archbishop of New York, made the brave comment: 'The US should search its soul about whether its role in some parts of the world could have created a climate of hate'.

Yet unexpectedly through *perestroika*, President Gorbachev had come to meet the West halfway. He believed communism could change, but he overestimated the possibility of change in the Party. The communist barriers were breached, and the whole régime disintegrated. America and the West were caught unprepared, and the post-communist years have lurched from one situation to another. There has been much to be thankful for, but there was no directing vision to illuminate future objectives.

With many regarding the US as the world's policeman, and feeling the need of one, false assumptions about the role of a superpower quickly began to take root. Quite apart from charges of arrogance, it is hard to establish a partnership of equals if a preferred solution is already formed. The sanction of superior power has to be suspended for the purpose of genuine dialogue. There may well be a clear consensus that multilateral procedure is the only way forward, and for this there must be an enhancement of the role of the UN. But even pending such a development, international relations can be vastly improved. Democracy depends not so much on winning control as on the readiness to renounce it if popular consent is lost.

Power is something on which human nature has come to depend, and to renounce it goes against the grain. But in actual fact it becomes clear that, with increasing education, imposed solutions will have a shorter and shorter life. The failures of the war in Iraq are clearly to be seen, but simply to be anti-war now is equally misguided. There is an opportunity for the Iraqi people to chart a new course, and we need to do our best to make it possible. In a democratic framework, when power is lost it has to be won back by raising your game rather than by vilifying your opponents. It is a time for hope rather than recrimination, and to redouble effort.

In the EU this is certainly such a moment, because public support has stalled for the want of a new vision. Although the recent enlargement (May 2004) which has taken place is bringing new voices into play. Europe is still a continent learning to be at peace with itself. That is the heritage of the post-war reconciliation, and the continuing challenge to extend it in the former Yugoslavia and further east. The new constitution may have failed to fulfil expectations, but it also had to deal with a way of getting business done. The extra powers for the European Parliament and qualified majority voting were the most important moves to this end. What may come out of the current turmoil remains to be seen, but it does not invalidate the shared values which have brought us this far. Europe's influence will stand or fall on the way it lives its faith, by whatever road its members have reached it. On the foundation of reconciliation and healing, goodwill can undoubtedly prevail. Moreover such a united regional voice will be a valuable and stabilising contribution to the world scene.

In Africa good governance is a growing theme, with an acceptance of responsibility for the way things are and for the changes needed. Although this is reflected very unevenly across the continent, an attack on corruption and the battle for clean elections have made headlines, which encourage expectations of progress. A recent resolution (February 2001) from the African Farmers Committee of the International Federation of

THE CURRENT POLITICAL CHALLENGE

Agricultural Producers emphasised the need to put agriculture back at the top of the development agenda. 'Agriculture remains the mainstay of African economics. Neglecting the sector that the majority of the population is involved in could spell disaster for many African countries.' There was also a call for African governments 'to harmonise policies and standardise trade regimes, to ensure political stability and to bring all stakeholders – including farmers – on board through consultations and sensitisation'. It is not at all clear that this call has been understood, still less acted on. Certainly outside Africa, there are too many who erroneously believe that international trade is the whole answer. Yet as Dr Christie Peacock, chief executive of Farm Africa pointed out (July 2004), 'There is a pathetic $2 billion worth of intra-regional trade in Africa, yet there is a $50 billion domestic demand for food crops, which is expected to double in the next fifteen years.'

In Asia these points seem to be better understood. India has benefited from lessening the stranglehold of government regulation, but is wisely preparing its own pattern of development rather than simply following the mainstream globalisation. A new centre for research into good governance promises to be at the heart of it. The vision it develops is important because, if there was only one pattern, we would be unlikely to bring humanity to the highest level of which it is capable. Indian democracy is now firmly rooted, and because it does not make big efforts to hide the warts from public view, is widely underestimated. There is a sense of responsibility and a readiness to listen to the inner voice which is the true heartbeat of a live democracy. India has much the West should listen to and note.

China has been opening itself to the outside world for some years now, and CAFIU (The Chinese Association for International Understanding) is doing a valuable work in developing links with other countries. Yet the greater part of the iceberg still lies hidden to most outsiders. For the leaders in China today it is becoming clearer that the great leap forward economically may end in tears, unless some moral purpose can supply a foundation to build on. They recognise that morality is the cement which holds society together, but it is not clear if it is understood that new standards cannot simply be established by collective means. Each individual has to find their own role, and that is something for which the Communist Party is not well adapted. Human nature requires a deeper understanding. Tiananmen Square may not yet be a subject which can be openly addressed, but the evolutionary political change needed is scarcely possible without it. Honesty is often painful, but it opens many doors.

Can progress towards democracy not be embraced with great benefits for the People's Republic? Or can a new morality be established without it being the fruit of a spiritual experience? Clearly these are live issues within the Chinese leadership, and many of them would like to see the harnessing of youthful ideals in voluntary service and initiatives of real social significance. Some hesitation is understandable, because democracy clearly signals a loss of control for those in power. But it could be seen as an opportunity to refocus with fresh clarity on what the country needs, and what people really want. At root it is a recognition that the Communist Party can no longer play God. But it does not mean that the People's Republic cannot forge its own pattern of globalisation, or contribute to fundamental changes in the current practice of capitalism.

Britain itself has had many lessons to learn in the last half-century, with the dissolution of Empire and the adaptation to changed circumstances. Harsh things have needed to be said about the way we always assumed ourselves to be right, without even stooping to self-defence. But changed attitudes have begun to take root. Some changes have been forced on us, some have been recognised as needed, and some still have to be accepted and tackled. But important new horizons have been opened up.

THE CURRENT POLITICAL CHALLENGE

Empire has given way to the Commonwealth, and though some decry it as more form than substance, the Commonwealth stands for certain shared values rather than physical power. It can claim to be one of the most egalitarian international organisations in the world today: the 53 members have an equal voice and Malta will be next in the Chair. It recently received a vote of confidence when the EU gave it a multi-million euro cash grant to help build trade capacity in developing countries.

But our association with the EU has been more ambivalent, and reflects our own uncertainties more than Europe's need. This may be why the relationship with the United States has been one of mutual endorsement rather than mutual change. So we still have to master the art of listening to truth from critics, while making an independent decision on what is right. It is as though we have abandoned the divine right to run the show, without fully taking up the teamwork needed to replace it.

In considering the opening up of new perspectives on the political future, it is worth reflecting on the moral and spiritual issues involved. Sonia Gandhi, in refusing the Indian premiership, spoke of listening to her inner voice to resolve her course of duty. She also said – 'I feel very strongly about India being a secular state. By secular, I mean one that will encompass all religions.' That raises the question of the part religion plays in human society, when, for all the secular spirit of the age, the quality of life and the attainment of happiness are still undefined in the public mind. So finally it is the purity of purpose which remains the key to renewal and renaissance, and that is the particular preoccupation of religion. Whether acknowledged or not it continues to be a formative influence.

9

Religion – the common heritage of mankind

Leo Tolstoy, who is celebrated as a giant of literature through his novels, is less well appreciated for the religious thinking which dominated the latter part of his life. Some feel he was confused and ineffective, yet Lenin dubbed him 'the landowner obsessed by Christ', and blamed his non-violence for the defeat of the first revolutionary campaign in 1905. Meanwhile his passionate attacks on the corruption in the Russian Orthodox Church and institutional religion only ended in excommunication. But he continued to proclaim that the common heritage of all religions was enshrined in universal and agreed principles.

> They are as follows: that there is a God who is the origin of everything; that there is an element of this divine in each person, which he can diminish or increase through his way of living; that in order for someone to increase this source he must suppress his passions and increase the love within himself; that the practical means of achieving this consist in doing to others as you would wish them to do to you.

This is a theme cherished by a high proportion of the human race. The Dalai Lama, Tenzin Gyatso, has written – 'Spirituality

I take to be concerned with the positive qualities of the human spirit – love, compassion, patience, forgiveness etc'. But in calling for a spiritual revolution, he underlines the need for high ethical standards to be part of it. He compares the undisciplined mind to an elephant which blunders about out of control. In Islam this discipline insists that the moral imperatives are a valuable resource in combating social and political injustice. But, contrary to popular opinion, it teaches an ethical position in which Muslims and non-Muslims enjoy the same rights, obligations and liberties. It understands that the moment religion is coerced, it breeds hypocrisy.

Yet, despite the widespread spiritual hunger today, there is an almost equal distrust of religious institutions as such. This is often explained by reference to those wars and divisions of the past, which were driven by extremes of bigoted dogma, fuelled by fanaticism. Although it is true that Christians managed to view the Crusades as a holy war, Muhammad himself had warned: 'Beware of extremism in your religion, as people before you were destroyed themselves because of their extremism.' While today, Tony Blair has drawn attention to the way people misuse religion to cloak sectarian strife. Speaking after September 11, 2001, he said,

> I'd like to start by making one thing absolutely clear: what happened in America was not the work of Islamic terrorists. It was not the work of Muslim terrorists. It was the work of terrorists, pure and simple. We must not honour them with any misguided religious justification... Blaming Islam is as ludicrous as blaming Christianity for loyalist attacks on Catholics, or nationalist attacks on Protestants in Northern Ireland.

All the same in both cases religious bigots have inevitably lent themselves to stirring the pot.

Despite this reaction to institutional religion which is too often misdirected, there are many who believe that the twenty-first century will see a global religious resurgence in both private and political life. This is based mainly on the conviction that religious pluralism is a fundamental resource which can be tapped by contemporary society to establish peace and justice. But it should be clear that religious pluralism calls for an active engagement with other faiths, rather than a passive tolerance. There is a need not only to tolerate but to understand, which means real care and listening.

This introduces a new factor into evolution, with which as yet we have hardly come to terms. To know how things happen is not the same as knowing why they happen. Richard Dawkins may be right to castigate us for not taking more precautions against the tsunami waves, but people are still troubled as to why our world should function the way it does. Professor Swaminathan has pointed out that mangrove forests protect the coastline from the worst effects of such events, but somewhere human nature has not been equal to making the most of existing knowledge and observation. We are struggling to bring good environmental practice into modern living, but even with it we are not going to bypass every crisis or potential disaster. The human spirit requires a rather deeper study of our destiny, even if our vision remains incomplete.

What religion brings to the question is a wealth of thought and experience which cannot be ignored, even if it fails to prove the unprovable. It speaks through silence and reflection, and concepts of God can grow in many different ways, once we are open to make an experiment with the thoughts or direction which come to us. It is essentially a living thing although, as a spirit, science cannot measure it biologically. That is perhaps why the earliest religion began with Animism, and why today the aboriginal and indigenous peoples have the sense of a great or powerful spirit in nature, deeply embedded in their culture. Perhaps, too, when farming led to the development of civilisations, it was natural that farmers should continue through thousands of years to believe in a creator, whose hand is reflected in the many wonders of nature and the ordered rhythm of the seasons.

The Dalai Lama, though not himself a believer in God, suggests as a Buddhist that, despite other ethical guides, the valuable human qualities are most easily and effectively cultivated within the context of religious practice. He observes that people who have developed a firm faith are much better at coping with adversity because they draw on an inner strength. There is no need for buildings or complicated philosophy, because our own hearts and minds are the temple or the laboratory where new qualities are refined. But whether we conceive a belief in terms of God or conscience, there must be an authority beyond that of the State or human leadership.

It would seem to follow that in any predicament there is a need to define what is right, rather than who is right. But different people arrive at different answers with an equal sincerity. Yet a valid consensus can be built by those who share a common foundation, and in religion this will be associated with God's will and a divine purpose for the world, which can only be developed by insight and the way we live. So it is on the practical level that we can come together, and find common cause in meeting world needs.

Professor Sachedina, an Islamic scholar, explains that Islam is not exclusive, but recognises that total surrender to God, or commitment to his will, is the acid test for everyone. It is the test of the decisions we make every day. This means the choice between doing what we feel like, or actively seeking God's purpose. For Christians the same sentiment is captured in the observation that 'Not everyone needs to become a Christian, but everyone needs an experience of the Cross'. Such a degree of selflessness is the only guarantee of humility, and the recognition that no one is in possession of the whole truth. Rather we are on a pilgrimage together, and the honesty of our living has to validate the ideas which we may feel have been revealed to us. Yet this moral understanding, implanted in everyone, allows and even calls for the development of a global ethic.

At every step this must leave us open to change, something which is rapidly becoming the hallmark of our times. But the starting point has to be change in human nature, not some avoidance of such a challenge. Charting a course without a moral compass is certain to complicate rather than simplify the task. As a recent study of St Paul's strategy suggests,

> Selfish men cannot create a free society, bitter men cannot build a lasting peace, morally defeated men cannot rise above self-interest. Those are inner moral contradictions by reason of which materialism as a way of life has failed humanity all through history and must continue to fail.

But Paul's life and work reflect his discovery that a radical change in human nature was possible. His immense energy and passionate conviction found a new direction in a faith stretched to include everyone.

For one thing, he saw that without a commitment to changing human attitudes and reshaping history, a person's basic motives are untouched and self-will unbroken. 'He didn't peddle a movement, but he brought the Cross to bear on the basic motives, entrenched viewpoints and cherished habits of man. He knew you cannot think relevantly unless you live adequately.' How you think certainly has a link to how you live. I have found for myself that farming decisions can be skewed, if I am not completely honest with colleagues at work or with my wife at home. In that sense Paul is a man of the future rather than of the past. A man who looked for a change in human nature to move forward the evolution of mankind on a global scale, and it makes no difference that his known world at that time was a fraction of what in fact it turned out to be. The truth on that point has expanded, but the truth about human nature remains essentially unchanged. We do not need to detach ourselves from our humanity, but we do need an evolution in character.

In the light of this we may need to rethink current beliefs about the source of real creativity, and the measure of contemporary progress. Sin is a subject which has been largely abolished in the secular world as part of the deal to exclude religion, but there has never been a keener search to find a sinner when things go wrong and people look for someone to blame. That, at least, is the public image, but in private life a good deal of liberal opinion feels people should not be made to feel guilty and can simply write off their failures. But while sin can be defined as transgression against accepted rules such as the Ten Commandments, it can also be seen as anything which separates one from God or from other people. St James puts the point well. 'You show me your faith without any deeds, and I will show you by my deeds what faith is... Whoever then knows what it is right to do and does not do it, that is a sin for him.' And the Qu'ran warns against the arrogance which leads to trampling on the rights of others – 'either they are your brothers in religion or your equals in creation.' Furthermore it describes repentance as writing a contract with the future; something more than just feeling remorse about wrongdoing

Inner freedom is fertile ground for creative thought, but it is produced by putting right what's wrong rather than shrugging off responsibility or seeking relief from the challenge of absolute standards. J.D. Unwin wrote in his classic *Sex and Culture*, 'Sometimes man has been heard to declare that he wishes both to enjoy the advantages of high culture and to abolish continence. Any human society is free to choose either to display great energy or to enjoy sexual license. The evidence is that it cannot do both for more than one generation.' So if a society pawns its freedom for the indulgence of its appetites, it takes a creative minority to bring it back on course. It should follow that if mankind accepts responsibility for shaping life on our planet, the relevance of such decisions becomes increasingly clear. Just as we are all, in Andrew Brown's phrase, 'interested in moral philosophy because we do it every day', we should also become involved in the ends for which we strive. It could be seen as both a religious duty and a political necessity.

History clearly provides much background on the evolution of human nature, and the value of cherishing a long-term purpose or sense of destiny. There may be apparently big gaps between generations but they are more easily bridged by honesty than by the belief that some inexplicable misunderstanding has been opened up. Still less that such a gap will somehow fundamentally change our perception of God. In a brief allusion to his own faith (*The Lexus and the Olive Tree*), Tom Friedman describes himself as a post-biblical Jew. I had not heard the phrase before, but he explained it as meaning you didn't believe God intervened in the world except when man requested it. This would seem to be worth greater amplification, which was not however forthcoming. Nor was there further reference to the values which the Olive Tree might contribute. In the Bible God's recorded intervention is both solicited and unsolicited, but in the modern world it is no less manifest. One person's perception of God's hand at work may be another's coincidence, but quite often there is a chain reaction caused by one person's change of heart. Human motivation comes from many sources, but God undoubtedly remains one of the most copious springs. The Olive Tree may have been a silent witness to a thousand years of development, but its contribution to the evolution of mankind seems to have been written out of the script. Yet to write post-biblical or post-Christian or even more modestly post-industrial or post-modern is not to close a chapter, and relegate a subject to the past. There is no convenient pigeonhole when talking about evolution and human destiny. That destiny continues to tease the imagination. So it may be a mistake to talk about post-this or post-that, when we could still need every speck of past wisdom to fuel our rocket into space. It may be a long story, but the sudden acceleration of knowledge should not entail the loss of our sense of direction.

Most notably this loss of direction can be seen in the failure to deal adequately with HIV/Aids, which has killed as many millions as destructive modern wars. There were some who forecast the scale of the danger, but much of the argument about dealing with it has been preoccupied with making condoms and antiretroviral drugs both available and affordable. Important as these may be, it remains true that prevention is better than cure, and in this case the cure has yet to be worked out. So the reluctance to face up to the consequences of promiscuity and sexual indulgence, which spread Aids, is particularly perverse. A recent headline in the *Daily Telegraph* (13 July 2004) stated, 'Love and trust will beat Aids, Museveni tells conference'. In the report President Museveni is quoted as saying, 'Aids is mainly a moral, social and economic problem. I look on condoms as an improvisation, not a solution.' Condoms represented 'institutionalised mistrust', and he preferred to look to 'optimal relationships based on love and trust'. The national campaign in Uganda is said to have lowered the infection rate from 30 per cent to 6 per cent. But whatever the figure, all are agreed there has been substantial progress. Yet the editorial in the same edition of the paper commented that Museveni had stressed the importance of concentrating on condoms.

This inability to face the objective truth and to decry solutions which call for a change in human behaviour appears to be widespread. What is more, it prevents beneficial changes receiving the attention they deserve. Some even like to hold the Catholic Church and Pope John Paul II responsible for deaths from Aids, when, if its teaching had been followed, the threat today would be greatly diminished. Talk of abstinence as the only alternative to condoms ignores Museveni's call for relationships based on love and trust, which are long-term. Those who talk of doing what they like as long as it doesn't hurt others are wildly unrealistic. They ignore the fact that when a relationship is severed unilaterally someone is always hurt. It may not involve violence, but to pretend hurt feelings don't count is a denial of humanity. Unhappily it has become the characteristic hypocrisy of our times. The UK is well down the European table on Aids, and could learn from President Museveni's good sense. For annual abortions in the UK have crept up to 185,000, and no amount of condoms will assuage the human suffering behind such figures.

It is this quest for dealing realistically with human nature which no doubt prompted Professor Sachedina to write, 'I want to go beyond the theology of inter faith relations to the theology of international relations for the twenty-first century.' He traces the Muslim attitude to such a development as being based on the concept of co-existence, which was an essential element in Muhammad's view of Islam as a middle way.

> Without restoring the principle of co-existence, Muslims will not be able to recapture the spirit of early civil society under the prophet. Only later did the religious mission become obscured by the political vision which had unified lands from the Nile to the Oxus.

This offers an historic opportunity to share rather than preempt the agenda for the future. Jihad in its ethical definition (and it has no military meaning in the Qu'ran) is the struggle for a man to be reconciled with his conscience. It is part of the human struggle to establish a moral order on earth. Islam's overlapping social and religious ideals can inspire the creation of pluralistic, democratic institutions.

Such thoughts could be bracketed with Robin Cook's desire to see an ethical dimension in foreign policy, as pointers to fresh practice rather than utopian ideals. In his introduction to *Religion: the Missing Dimension of Statecraft*, Douglas Johnston lays out such a scenario, though more in terms of NGO initiatives than institutional policy. He is executive vice-president of the Centre for Strategic and International Studies in Washington.

> I concluded, rightly or wrongly, that the rigorous separation of Church and State in the United States, had so relegated religion to the realm of the personal, that it left many of us insensitive to the extent to which religion and politics intertwine in much of the rest of the world. Such an insensitivity, I further speculated, could lead, and probably had led, to misinformed and potentially costly foreign policy choices.

The book itself examines hard evidence of what has actually happened. The chapter by Edward Luttwak on Franco-German reconciliation after World War II evaluates the role of Moral Re-Armament (MRA), now known as Initiatives of Change. He notes that, despite slender resources, MRA was in the field at the right moment and with a committed team, even while Germany was still occupied and travel outside depended on the respective Occupation authorities. He sums up his researches by saying,

> As far as MRA's parallel diplomacy is concerned, the evidence cited here is literally conclusive. Its contribution was clearly limited and just as clearly of proven significance: MRA did not invent the Schuman Plan, but it facilitated its realisation from the start.

Among other situations documented in the book is the role of the Quakers during the Biafran Conflict in Nigeria. Arnold Smith, the Commonwealth General Secretary at the time, was asked whether the Quakers were seen as religious or secular actors. He replied, 'I think people who knew them understood that their motivation was spiritual. But spiritual is not an opposite to secular, it's an attitude of values, of how you deal in secular matters.' This puts in a nutshell the point that a new and selfless attitude can help in bridge building, but it should not disguise the fact that trust will depend on people who clearly live by the standards they advocate.

This book, published in 1994, establishes a fresh line of thought, and a study of its contents would throw much light on what many countries sincerely hope for from Washington policy making. Track 2 diplomacy, as it has been labelled, clearly has a continuing and growing role to play in an increasingly democratic world. What NGOs have done is to proclaim a disinterested purpose, and provide an opening for popular participation. Indeed Voluntary Service Overseas in the UK in seeking recruits advertises 'a lifechanging experience', and it is intriguing to speculate what that may mean for different people. It certainly implies a spiritual aim in the sense of caring for other people, and idealism in the sense of putting good ideas into practice.

A pioneer programme known as Action for Life (Initiatives of Change) has taken this concept a good deal further in a nine-month programme of training and travelling. From November 2003 to July 2004 forty people from twenty countries, of different faiths and backgrounds, came together in India and later moved out in smaller groups across Asia. Four points summed up the basis for their teamwork:

> Change yourself.
> Engage others.
> Create answers.
> Give hope to humanity.

Looking back on their journey, the headlines were:

> Seeking better governance in the largest democracy in the world.

> Experiencing the after-effects of change in India's heartland.
>
> Holding on to spiritual and cultural heritage in a rapidly changing society.
>
> Cherishing the riches of diversity in the move towards a global community.
>
> Understanding the role of the individual in Asia's economic and human powerhouses.
>
> Looking beyond history's wounds to horizons of hope.

Their report is worth studying on www.afl.iofc.org and 2005 saw a further chapter. It all flowed from the 'crazy idea' of Ren-Jou Liu, a teacher from Taiwan.

He saw it as a way 'to mobilise a new generation of change-makers equipped with integrity, faith and commitment, and dedicated to bringing transformation, healing and development in Asia and the world'. An Asian-based initiative which welcomes the West, it could become a pathfinder in shaping the future.

Such a programme is providing direction for young people who are seeking it, but have not had the opportunity to see how to make it practical. It is at once an individual calling and a demonstration of the power of teamwork in producing a new spirit. The inspiration has to come from within, so a purely secular approach is hardly possible. The vision of a new world has to be constantly renewed by faith, and live in the lives of those who experience the leading of the spirit. But it remains a universal heritage.

10

God is His own interpreter

> Blind unbelief is sure to err
> And scan his work in vain,
> God is his own interpreter
> And he will make it plain.
>
> William Cowper 1731–1800

Cowper's verse captures perhaps the kind of evolution we need, from the accumulated records of institutional development and the definitions of doctrine to the daily experience of the promptings of the spirit. It is the discovery of a God without labels, who can lead us into fresh understanding, and the perception of a world vision. He is always there for communication, and will give truth to everyone who cultivates the will to listen. What is more, there is no need to believe in order to begin, and everyone will have their own starting point. As Cowper expresses it, God has it all worked out for us including human helpers.

One starting point is our own need, which may or may not seem pressing to us. Another, the needs of the world, which without doubt are pressing if we open our hearts to them. Whether it is the Indian Ocean tsunami disaster, or widespread poverty, or ruthless governments, we are looking for global answers. If they cannot be achieved overnight, we must at least be able to see a potential pattern of evolutionary change. It is that which turns us towards the reading of history, and an attempt to decipher the formative influences at work. Father Alphonse Gratry, a well-known French Oratorian priest of the nineteenth century, wrote that the study of history soon led to three questions. First – where do we stand in relation to our ultimate goal? Second – by what route have we arrived at where we are? Third – what is the journey still to be accomplished, and what can the past teach us about the future?

He remarks that the first point is never addressed, although it is readily seized on by political commentators.

It may be a point on which some progress has been made. Gratry was a man who trained his students to take the measure of the century in which they lived, and to dedicate their lives both to expressing and living out the answers to the problems they encountered. For him this involved learning to make God their tutor, and writing down the thoughts which came in times of quiet reflection and meditation. He saw his own task as reuniting religion and science, which he believed to be a key challenge facing the nineteenth century. Living in the wake of the French Revolution and the Enlightenment, he was nevertheless able to welcome new thinking, without feeling that it invalidated the experience of faith.

Gratry would have appreciated the inscription on the grave of Karl Marx. 'The philosophers have only interpreted the world in various ways – the point is to change it.' But he would have been well aware why Marx had misread the nature of religion. Communism became the god that failed, because it never got to grips with human nature. Although it claimed to be the gospel needed, its doctrine became the ultimate example of laying down what was politically correct. Yet Gratry was also the architect of a vision for our times when he articulated the need for a universal commitment to ending poverty and misery, as a sign that mankind was going to enter a new era. More surprisingly, he also forecast that one of the next big discoveries would be in the realm of atomic science, and that one ought not to put limits on asking questions such as those about the potential of the atom. Was he wrong to suggest, in reviewing comparative science, that the road should run through the mathematics of physics and chemistry to the realm of philosophy and theology?

Interestingly enough, Father Gratry describes theology as philosophy which has been grafted with the spirit. I take this to mean that there is a point at which faith and intuition enter in, to shape our ideas and mould our thinking about the purpose of life. This may not be quite the same as the experience which transforms people, but it is part of making ordered sense of the inner workings of our spirit. It leads away from the kind of devotion to dogma which can be used as a club to beat people over the head, to the pursuit of truth to the point where I can truly say that my life is given, and is no longer my own. But for myself, and perhaps for a majority, it is difficult to define this experience in more revealing words. Yet, as Marx believed, the point is not just to explain the world, but to change it. Religious rivals, and even enemies, can be reconciled like any others when an inward truth takes over. A notable and recent example of this kind of reconciliation comes from northern Nigeria, where conflict between Muslims and Christians sometimes boils over into violence. Muhammed Nurayn Ashafa, the Imam of Kaduna, and the Revd James Movel Wuye are joint directors of the Inter Faith Mediation Centre in that city. Their story is taken from the magazine *For a Change* (see Preface)

'We were two militant religious activists,' said the Imam, 'but now we are working to create space, not just for peace, but also for the transformation of society.'

Movel Wuye continued, 'We were programmed to hate one another, to Islamise or evangelise at all costs. This threatens the very existence of Nigeria.'

'We were both victims of the situation that we had both had a part in creating,' Nurayn Ashafa added. His spiritual master and two brothers had been killed by Christian militias; Movel Wuye had lost an arm in the violence.

'What motivated us to transform hate into love, vengeance into reconciliation?' the Imam asked. 'In our hearts, we were weeping, but we were still full of hate.' A turning point had come when he heard another Imam preaching in the mosque at Friday prayers about the power of forgiveness, and the example of the Prophet. This had led to a war within, he said. Then embracing Movel Wuye beside him, he added, 'he is no more an enemy but a friend.'

Movel Wuye said that it had taken him three years to overcome his hatred and to start to trust the Imam. The process had started when Nurayn Ashafa had visited him after his mother died. An American evangelist had told him that you cannot preach to someone you hate; 'He was radiating love, but I'd been blinded by hate and pain,' he added. Now they were working with other spiritual leaders, 'to create space for peace and understanding'.

This experience reflects a return from the wilderness to the paths of true religion. It has already set in motion a completely new intercommunal dialogue, and may yet prove an historical turning point. Certainly a great deal more will be heard about it.

It mirrors the experience of St Paul on the road to Damascus, when as a devout Pharisee he faced the fact that his persecution of the Christians was misguided. No doubt his presence at the death of Stephen by stoning was a formative experience which worked within him but his stunning conversion on the road did not provide him with a ready-made path. It was followed by fourteen active but unrecorded years before he eventually set out on his travels with Barnabas. He was led at every step as a man under command, who sought to do God's will. He proclaimed the worth and equality of every soul in the sight of God. He could see that ultimately he would reach Rome, the heart of the superpower of his day, as a prisoner, but that a whole new process would be set in motion.

In a foreword to Paul Campbell and Peter Howard's booklet on *The Strategy of St Paul*, Klaus Bockmuehl, lately Professor of Theology at Regent College, Vancouver, wrote, 'In refreshingly down to earth contemporary language they express St Paul's world view for our own age, in a way which could help us rediscover the almost forgotten art of interpreting world events from a Christian standpoint'. Such a standpoint is non-sectarian and universal. St Paul, himself, fought strenuously against labels and personality cults, declaring that he had no monopoly of the truth which belonged to everyone, whether Jew or Gentile. Yet his testimony was unswerving, that above all he worked to fulfil the purpose for which Jesus lived and died. He had a clear priority – 'this one thing I do'. Armed with this commitment, he found fresh significance in the daily round and its potential outreach. H.G. Wells, who was not himself a Christian, suggested that 'the test of greatness is what a man leaves to grow. By this criterion, Jesus stands first.' It is certainly true that Jesus is welcomed in many parts of the world, where Christianity may be seen as at best a mixed blessing. Christian institutions have collected a lot of baggage through the years, but it should not be allowed to limit today's vision. There is room for a fresh understanding of what Jesus actually set in motion, and what it means for the world today.

Most strikingly, perhaps, he was ready to engage with everyone he met, and never questioned their antecedents or loyalties. His message respected tradition, but cut through hypocrisy or humbug which sought to cloak moral compromise with respectability.

Yet he was always there to meet people at their point of need, whether it was of their own making or not. It might be quite hard to think of him promoting Christianity as a religion, when his reference point was always to God the Father. For Christians he remains the supreme incarnation of God's power, but he always recognised the validity of prophets and saints through the ages. The people of the Book (Jews, Christians and Muslims) remained in some degree colleagues on the road, even if they were not aware at that point of Hinduism and Buddhism and the widely diffused spiritual traditions of the East. The latter are the source of much inspired and universally acknowledged thinking. Although sometimes puzzling to the Western mind, they are now part of our global understanding of the spiritual evolution of humanity.

History is the story of human struggle, but for those who tend to see it in terms of conflict it is hard to come to terms with the violence of modern war. Yet General MacArthur probably thought he was stating the obvious when he said that the springs of human conflict cannot be eradicated through institutions, but only through the reform of the human being.

Tolstoy supposed that the deaths dealt out in hand-to-hand combat always had their effect on the combatants, whereas today missiles are delivered from such a distance that the personal link is broken. Television makes it graphic, but even so the constant repetition of such scenes can dull the sense, or make one take refuge in blaming the leaders who have let it loose. But I can remember as a boy listening to a radio report on the war between China and Japan in the 1930s and seeing in Dr Frank Buchman, the initiator of Moral Re-Armament, who was staying with the family, someone personally involved through his friends in both countries, who felt the pain of the battle being described. It made a deep impression on me because Frank Buchman knew some of the leaders, and was a man of prophetic foresight, who was committed to international change. That he shared the suffering described became clear to me from his comments. At some time before his death in 1961, he remarked to a younger man, 'It's the best life in the world. I would choose no other. But I would not like to live through the times you will have to face.' Yet he foresaw the end of communism in the lifetime of those around him, and warned them they would have to be ready to meet the needs of those whose society had been destroyed.

That is why a wholly new perspective may need to be discovered today, and why a conscious pursuit of the evolution of man may have more to do with a new spirit than the creation of wealth or the growth of knowledge. To seek God's direction is the path of destiny for those with faith. It is also a path open to those without faith, if they wish to make the experiment. But to conceive a human destiny through a collective body rather than the individual has been decisively rejected. Such technological societies have been conjured up as horror stories by Aldous Huxley and George Orwell, quite apart from real-life examples. So it follows that the individual democrat must rise to new heights if evolution is to follow a benign path.

For if man is in charge of his own destiny, there is need for more than human wisdom. Many practical materialists are ready to acknowledge the importance of the spiritual, even though they have no intention of letting it change their course. I have noticed that letters questioning vision or motivation are frequently answered with comments about them being 'insightful'. Sometimes it is no more than good public relations: an attempt to improve a company's reputation and live up to some expressed mission. But it can also be an acknowledgement that present proceedings require a moral and spiritual dimension, which has been excluded.

As John Bunyan made clear, we are on a pilgrimage in this world, which cannot be complete until our death. For those who approach death in full consciousness of its coming, it is often the meeting point of faith and love. A strong faith drives people on through every battle, but it is love which at the end takes charge and is beyond all understanding. The intimations of immortality may come to us at varied points in our lives, but it is mostly with the advance of old age that we feel prepared in spirit. That is an inward journey, a perception or a gift, which comes to us without asking and cannot be shaped to our own purposes. A recognition that, if there is a God at work, that's him. Near-death experiences attest to the compassion with which we shall be received, but nothing can anticipate the final moment or leave more than a partial reflection with those left behind.

Yet for all the gifts one may be given along the way, making the decision to embark on a pilgrimage requires a certain courage. Indeed C.S. Lewis has suggested that 'Courage is not simply one of the virtues, but the form of every virtue at the testing point'. By this I think he meant that courage is needed to implement every step in change, whether it involves humbling our pride, taking a risk for the benefit of others, or denying a leading passion of the moment. Even love calls for courage in its giving, and is met by the grace to fulfil our task, and to show virtues we did not know we possessed. It may be an important truth on which to reflect, at a moment when it is clear we face the unknown and have to make the experiment of personal experience, which requires persistence and perseverance. It is also something on which we have to embark alone and in silence. To achieve that silence there may be battles to be won to break the slavery to self or to the accepted values of our age. As Siegfried Sassoon wrote,

> Alone…the word is life endured and known,
> It is the stillness where our spirits walk,
> And all but inmost faith is overthrown.

Yet his words seem to celebrate the test rather than the victory, because that is the moment we realise we are never alone or without God. If God is his own interpreter, it is clear that he tells those who really want to know. Those who seek are those who find, but they have to realise that it is a long road. Truth is like a cut diamond which has many facets, and shines from many angles. Where one perception leads to another, there can be no turning back. When Jesus sent his disciples out ahead of him, he underlined the depth of commitment required. 'No one is any use to the reign of God, who puts his hand to the plough and then looks behind him' (Moffat translation). But some may have already paid a high price, as the experience of Dr Yusuf Omar al-Azhari illustrates: it is a striking example which I reproduce from the magazine *For a Change* (see Preface).

> As an ambassador and leader of my country, the future was golden. To complete the picture, I thought the thing to do was to marry the daughter of the President of the country, which I did. The future seemed assured.
>
> Suddenly, in a coup, my father-in-law was assassinated and a military junta took over the country. Everything began to change. A year later I was picked up at home in Mogadishu at 3am, handcuffed, blindfolded and driven 350 kilometres where I was put into a cell 3 x 4 metres. I remained in solitary confinement with nothing to read, no one to talk to for six years. During the first six months I was tortured daily. Possessed by anger and hate, I hit out violently wherever and whenever I could. I was afraid I would go insane or die. My brain was trying to burst.
>
> One night in my desperation, I knelt down to pray at 8pm. I asked the Almighty for peace within me and for a purpose to guide me. When I finally got up from my knees it was 4am. The eight hours had passed like eight minutes. I had never had a better eight hours of prayer in my life and I felt exalted. I felt cured and free of hate, despair and depression. My desire for greed and enjoyment had gone. I felt personally accountable for all my actions.
>
> The guards were surprised to find me calm and submissive the next morning. They could no longer torture

me as they had done before when I had reacted to them. They wondered how I could have changed overnight. Having decided to accept prison life, I divided the hours between physical exercise and conducting debates with myself about the past. I would spend hours thinking back over the wrongs I had done. But also I traced back the good things, so that I did not get obsessed. Six years in prison gave me a training for life which I could not have received in any other way.

In 1991 the military rulers who had taken over the country were deposed and the Commander-in-Chief, Major-General Mohammed Siad Barre, fled to Nigeria, where he was given asylum. When I emerged from prison, I went in search of my family and found them living in a hut. When my wife saw me standing there, she fainted. She had been told that I had been shot trying to escape. She had not realised I had been imprisoned.

'Can you forgive the man who has done all that to you?' was the question I kept asking myself. One day sitting in a coffee shop, I had the strongest feeling that I should forgive the man who had caused me such misery. But it took me two years to decide to do it. How to get a ticket to Nigeria when my bank account, my house and everything had been confiscated? I had no money. Three days after deciding to forgive, I was asked to represent Somalia in a UN conference in West Africa. This enabled me to visit the former dictator, now 87 years old. I went all that way to tell him that I forgave him. As I spoke, I saw the tears roll down his cheeks. I thanked God for giving me the chance to fill such a man with remorse. He said to me, 'Thank you. You have cured me. I can sleep tonight knowing that people like you exist in Somalia.'

Since that experience, I have been working without an official position to bring peace and reconciliation to my country. We had no government, no judiciary, no police force, no schools. My experience of forgiveness gave me self-confidence and the realisation that I could befriend anyone fearlessly. This became my way

> of life. I found it is the honest way to win the hearts of others.
>
> Ten years ago with the help of Swedish facilitators, I found the way to build friendship and trust with other Somalis through my readiness to apologise for my mistakes and my hatred. We are beginning to find a way forward to rebuild our country and recreate the administration. Through Allah, we peacemakers conceived this new approach to nation building.

The eight hours he spent on his knees in the cell are the measure of a new world being born, not only in one man, but in an experience which has reached thousands in many nations. It transcends every human barrier, and embodies what is needed to find a new road. Those who see the future in terms of military or economic supremacy may be discussing some of the pressing issues of our time, but they may still be asking the wrong questions. To measure the potential of the coming century is a long-term task, and there are many voices striving to be heard. Perhaps in the end we may have to return to silence, and a totally fresh perspective, where God is his own interpreter, and where life takes on new meaning.

A Damascus experience may astonish many observers, but it only becomes fully understood by those who choose to follow the same path. Those who begin to find that their life is no longer their own. Henry Drummond used to warn his students that such sudden discoveries were the exception rather than the rule. Quite radical decisions soon seem mundane. For most the next day will be very like the previous one, but looking back in time the consequence of such decisions seem immense. So although it may or may not be dramatic in human terms, if God is working his purpose out, we can only understand it through the commitment we make. That is what counts in the cause of a conscious evolution, which offers every individual their unique part.

PART II

The farmer's contribution

11

Why farming is different

I have made a brief tour of science, economics, politics and religion because all have major influence on the future and the farming scene. To me they are the fundamental keys to the change of direction that is needed worldwide, to which farming has a great deal to contribute. The divorce which seems to have taken place between town and country in the UK is much less pronounced across Europe as a whole. Nevertheless Western urban society assumes that agriculture is a secondary activity in a sophisticated world, and that a high-tech future can continue to cut costs and deliver cheap food. So it is vital that some of the realities should be better understood, and that farmers should succeed in articulating what they would like to bring to a future which is sympathetic to people, and respectful of natural evolution. It is not an industry like any other, because it involves a way of life which has many ramifications.

Farming, in fact, is about life, work, family and our relationship with nature. It holds the key to two of the big issues of our time – closing the gap between rich and poor, and the long-term stewardship of our environment. So although it has come to be called a business like any other, it has always been a business like no other, in the sense of its outreach into the culture of our society. In the West, farming has been relegated to a minority role to which no great importance is attached. Only recently have the public become aware of developments, which they consider undesirable. But in the developing countries farming is the life of the majority, and is the key to answering poverty. No other industry or activity can match agriculture as a source of work, and that still holds good. The onward march of technology may continue to produce more with less effort, but it will not in the foreseeable future provide all the jobs and employment that people are looking for.

The divorce between town and country as an issue has been increasingly forcing itself on the public mind, and many are now addressing it. It came home to me afresh when I wrote in an article at the time of the Foot and Mouth outbreak that the government's brief 'shutdown of the countryside' showed why farming was different from other industries. This seemed to be regarded as a non sequitur, and it was pointed out that the tourist industry generated more money than agriculture. What did not appear to be understood was that the countryside, which attracted ramblers, fishermen, horse riders and other lovers of rural life, was a product of farming. The whole mosaic of crops, livestock, hedges and stone walls reflects a human creation. Agriculture is in fact also an integral part of the tourist industry, being involved in bed and breakfast, caravans and camping, farm shops, farms open to the public, and much more. Over half the farmers in the UK already have some source of income outside food production, which reflects how much the rural economy has already changed amid the pressures to survive.

Opening farms to the public and particularly to schoolchildren is already making it clear that milk comes from a cow rather than a supermarket. But some of the other misconceptions take rather longer to dispel. At the extreme end of the spectrum are those who would claim that we scarcely need any productive agriculture, and that food security is no longer a factor of real weight in farming policy. Yet three times the shipping currently employed would be needed to implement such a policy, and it is certain that the disruption caused to world production would ensure shortages elsewhere. The value of the international food trade is easily exaggerated as well, when only about one fifth of production is traded on the world market. Moreover, in Europe, 89 per cent of trade is intra-regional, and that is a sensible pattern, having in mind the energy involved in simply moving goods around.

When it comes to the market, the great difference between farming and manufacturing is the fact that in farming production still depends on the weather, and there can be great fluctuations which are beyond the farmers' control. Moreover the farmer's land is both his capital and his workshop. He does not have the flexibility of downsizing during an economic recession, and then gearing up again when things improve. Self-employed farmers in the UK cannot pay themselves a living wage before

Index of Farming Income* in Real Terms (Average 1940–69 = 100)

1940–49	101	1990–97	66
1950–59	100	1998	25
1960–69	99	1999	23
1970–79	101	2000	15
1980–89	55	2001	17 (provisional)

* The return to farmers and spouses for their labour, management, own land and own capital, after providing for depreciation.

As this figure is no longer published, the figures from 1997 onwards are only approximate.

(John Nix, Wye College, London University)

calculating their profit, and the same profit figure has to cover capital investment in buildings or equipment. The levels of profit over the last 40 years, shown in the table, reflect a situation which no wage earner would tolerate. It shows that when it comes to food production, markets do not perform and need regulation of a basic nature. That is why agricultural subsidies have crept in, and have become an integral part of Western policy.

One thing which can be said, however, is that farmers have greatly improved their performance in accounting. In today's economic climate, those who don't accurately know their cost of production will not last long. Yet lowering costs will not automatically generate a margin of profit. Farmers undoubtedly need to exert more control over market prices through co-operation, but it is no simple matter. The factory farming which has become widespread in pig and poultry production illustrates what can easily become a race to the bottom – mass production with a wafer-thin margin per bird or pig. As the process has intensified, public doubts have been raised both about meat quality and animal welfare. Some change has already begun, and in this respect the UK can claim to be among the leaders in a process which, though still larger-scale, is a return to more natural methods where the animals' interests are paramount. Large-scale egg production is increasingly moving to free circulation housing with perches, and access to outdoors, while as much as 30 per cent of pig production is now outdoors, though its practice is heavily dependent on suitable soil types and weather patterns. But those who consider that animal welfare should be judged entirely on what animals do in the wild need to remember that domestic animals may not always share the same predilections. They have become used to a softer life – easy feeding and less exercise. So not all parallels with the wild may hold good.

Factory farming is, of course, a global phenomenon so it is more accurately measured in terms of world prices, when it comes to international comparisons. Climatic differences mean that in one country the aim will be to keep the animals warm, and in another to keep them cool. But the basic variable production costs will still be labour and food, and there is no question that financial pressures have led to the more indefensible practices. Chief among these has been overcrowding, and the regular feeding of antibiotics. So it is an important development to find that many of those who have been critical of such things are ready to pay more for their meat if it is produced from healthy and good living conditions. This has led to what has become known as Farm Assured production, and the traceability of meat down to the individual farm and animal. But this is a long way from standard practice outside Europe.

When it comes to world prices it is commonly assumed that developing countries will always be producing more cheaply. This will be generally true because of the lower wages, but it does not ensure that their farmers will receive a proportionately more profitable price. If they did, wages could be expected to rise, but in general they conform to the worldwide picture where family farms normally rely on family labour, and large companies rely on manpower only so long as it can be had cheaply. Otherwise it must be large and expensive machinery operating over increasing areas. This means either huge farms or contractors serving a large area, and perhaps moving north or south (according to hemisphere) to extend their season of work for expensive combines etc. Such a pattern requires more analysis by those who pin their faith on increased international trade, because it is far from clear that major benefits will reach the front-line producers and workers. It is certainly a social as well as an economic issue, and one which is increasingly weighing on the environment.

Large-scale arable production may still have to convince the public that it is compatible with nature and biodiversity, but a considerable evolution has been taking place. In Europe a period of mono-cropping has now seen a widespread return to the benefits of rotation. Reducing chemical inputs of all sorts is recognised as a desirable objective, and accelerated by the need to deal with pollution and establish good water management. Changes in cultivation practices, particularly surface cultivation or direct drilling instead of ploughing, which may have been pushed at first as a cost-cutting exercise, are now seen as primarily to be judged on their contribution to soil conservation and fertility.

For those outside farming it is important to appreciate that growing clean, weed-free crops is bound to impact on biodiversity. It has been part of traditional good husbandry through the ages, and the assumption that all modern farming is inimical to wildlife is clearly mistaken. Much effort is being put into making waste corners of land good wildlife habitats, and attention is currently focused on the best treatment of field margins. More hedges are now being planted than removed, and many of these are for the benefit of wildlife rather than their traditional purpose as a stock-proof field boundary. But there must still be a distinction between farms and wildlife reserves. It may be good to preserve some wildflower meadows, but it is nonsense to suppose that they represent a desirable basis for grassland farming. They will in fact reflect a lower level of fertility, and good farming practice will aim at greater productivity. Then a larger number of cattle or sheep, allied to good grazing management, will promote a cycle of higher fertility. So there is a balance to be struck, and it is currently the subject of considerable debate. Yet farming, as nature itself, will always be a world of manifold solutions.

Still another unique aspect of farming is that the farmer lives on the job. His home and work are one. He doesn't need to ponder Ricardo Semler's conundrum – 'If you can take your work home, why shouldn't you take your family to work?' It is a real opportunity to develop family life, and make it still the heart of rural living. Family life has always been the core of Tony Blair's philosophy of community, and it is something for which he should be commended rather than criticised. Interestingly enough, one farmer friend of mine, who had decided against starting a family, changed his mind as a result of a visit to our twinning colleagues in rural Normandy. He was struck so forcibly by the joy of children on the farms there that he realised there was no better place for children to grow up. It is part of the pattern for a new world, rather than just something left over from the old. It is also one more reason why farmers worldwide share a bond and an understanding, which transcends language barriers.

That in itself may be a trump card in shaping the future, because it is globalisation with a human face. Affluence so easily erodes the qualities of hard work, self-discipline and community, if there is no vision beyond the present. So it is important that the character of an independent people remains or becomes the most important crop grown on farms worldwide. Then the members of the family, whose education takes them into many different jobs, will carry with them a commitment to values which shape a worthwhile society and bring unity in place of division.

For although farming was the forerunner of civilisations as we know them, the steady march of urbanisation has unconsciously led to the belief that cities are synonymous with civilisation. City life is seen as the end product, so to speak, as well as providing the more comfortable lifestyle of the better-paid, though slums and shanty towns give the lie to that for many at the fringes. For Westerners it is often a surprise to realise that farming is the majority activity on our planet. But not too many of them will think it likely that farming will remain a front-line activity. They simply project Western development on to the rest of the world with hardly a second thought. Yet climate change is constantly pushing environmental questions up the agenda, and food production remains a live issue at the heart of poverty reduction and coping with a rising population.

So long-held assumptions may yet be turned on their head by education and the advance of Information Technology. Asia may be the continent where such a change is crucial. It is only thirty years since the World Food Conference in Rome (1974) was called to consider the possible threat of a world food shortage. That is now a distant fear, but the fact remains that hundreds of millions are still malnourished. A revaluation of agriculture is unquestionably the key to change, and goes well beyond current arguments about trade and food quality.

Agriculture needs to be attracting the best scientific brains at the forefront of our thinking for the planet. Farming needs to be seen as the spearhead of our hopes to shape a new deal with renewable energy, with conservation of our soils' potential and with better water management. Nowhere is this more likely to be demonstrated than in Asia. Professor Swaminathan points to a growing chain of information villages in India which are developing direct links with researchers who have something to offer. He likes to say that he looks for the poorest woman in the village to be the one to operate the computer, because it immediately gives her status. That reflects the widespread Indian aptitude for computer technology, but it could also be a parable for our times, when the uneducated are more often planned for than included in making the plans. Swaminthan, himself, focused his ambitions for study on agriculture rather than medicine as a result of the Bengal famine. But he has never been the kind of boffin who simply hands out information from a laboratory. It has always been his purpose, as well as his pleasure, to get out on the land to meet the farmers and plan with them how fresh knowledge can be most useful.

The International Farmers' Dialogue, which is aiming to raise the profile of these issues, is only in its early stages. But it is to be hoped that the wider public can share in a reappraisal of their importance. It is not just about lobbying for a special interest group, but about the long-term future of our planet. Education, medical services and infrastructure are all vital ingredients of the development needed. They will surely accompany a reappraisal of agriculture not only as an economic powerhouse, but also a key factor in the stability of national life. A major change in the current direction of thinking and policy making is needed. It is needed not just to make agriculture prosperous, but to allow it to contribute what it is meant to. So it is the aim of the next chapters to outline this potential, and suggest that a revolutionary change in our thinking may be needed to put us on course for a desirable evolution.

12

The Common Agricultural Policy (EU) – the way forward

In any judgement made about the CAP, it is essential to take the long-term view. It is a significant policy enterprise, which should not be judged only by its failures or even disasters along the way. For those able to remember the early days, great strides have been made in mutual understanding, and more recently on the likely impact of the policy on other countries. For even the forecast of the benefits to be gained from the Uruguay Round, which led to the establishment of the World Trade Organisation, gave a minus value for Africa.

Looking back on the immediate post-war years, most people understand that agriculture was faced with food shortages, and feeding hungry people. Farmers felt themselves appreciated, and responded with a huge surge in production. Under the system devised, surplus stocks began to build up in intervention stores (intended to maintain market stability), and the failure to control them produced much bad publicity. Food mountains became the main image of the CAP, and there was much damaging evidence of the depressing effect on agriculture in developing countries, when attempts were made at their disposal. Some of this was due to government policy in the recipient countries, but it was a situation which took far too long to be rectified. In fact that process is not yet complete.

The first serious attempts to improve the situation led to a policy of supply management. It included individual farm quotas for milk, and set-aside for areas of land deliberately taken out of production. It put a ceiling on overall production, and this was reflected in the reduction of the EU's share of the global market. It also opened up a discussion of world markets, and the price levels at which countries should be expected to compete. Lower-cost producers may have a right to compete, but it is not much use to them if world prices still drop below their cost of production, which should include a fair figure for labour.

In 2005, a further important step has been taken, which separates the subsidy from production decisions, and outlines a concept of rural development in highly industrialised countries. The core of these new arrangements is that a single farm payment will be made to each farm individually, which no longer depends on what the land is producing. It will be based on an historic figure for what different farms have received in the past, and a small percentage (known as modulation) will be diverted for various purposes of rural development. The payment of this subsidy will be linked to certain environmental standards in the management of the land (known as cross-compliance).

The effect of these measures amounts to a decisive change in the direction of European policy. Farmers' production decisions will be directly linked to market prices, and any contracts they may be able to arrange. Subsidised exports will be eliminated, and the spotlight will be thrown on the performance of world markets across a whole range of commodities. It could prepare the way either for completely free trade, or more desirably for internationally managed markets, because farm subsidies have arisen from market failures. Reflecting the realities of world markets is clearly a commendable step in the right direction. But nobody should expect it to do more than highlight the need for a further change of direction, if farmers' livelihoods worldwide are to be secure.

Franz Fischler, the Commissioner for Agriculture in Brussels until the winter of 2004, is to be congratulated on seeing these changes finally under way. An avowed champion of the family farm, he has always understood that a balance had to be struck between the diminishing rural workforce and the growth of megacities. As he has commented,

> In discussing the economic arguments against supporting agriculture, no one ever seems to take account of the cost of providing under-utilised infrastructure in rural areas, while at the same time having to invest ever increasing amounts in cities to cope with urbanisation.

Such arguments are echoing afresh with the disagreement over Britain's rebate (June 2005). The fact is that, in terms of the national budgets of individual countries, the EU budget is a very minor expense. That agriculture should take nearly 40 per cent of it simply reflects the fact that it is the only comprehensive common policy there is. There is a big difference between harmonisation of national methods, and a single policy for the whole. It may be said that it is being spent on only 5 per cent of the people, but it involves a major portion of everyone's food and 100 per cent of the land outside the cities. From such a perspective it represents the changing face of a large part of the continent, and is particularly significant with the recent addition of ten more countries. The more so when the policy is now taking account of Europe's relationship to world agriculture. This may well be a case where facing the facts will shed much light on the realities, and their relevance for the future, because at present there are still huge misconceptions.

All this development of policy is contained in the history of the last fifty years. Yet on many practical issues change often seems painfully slow. Achieving a consensus takes time, and changing farm policies too fast can lead to dislocation and hardship for many individual farmers. What is more important is that the basic principles of policy can stand the test of time, and that the changes found to be necessary clearly move things in the desired direction.

In looking back at how things have evolved, it is quite hard to remember the insularity with which we emerged from the war. Although Sir James Turner (later Lord Netherthorpe), president of the National Farmers Union of England and Wales, was busy founding the International Federation of Agricultural Producers (IFAP), most farmers knew next to nothing about farming across the Channel. When I went to work for some months on a French farm in 1948, it was a major voyage of discovery. Despite the many differences in outlook, it cemented a sense of comradeship which has endured through the years. Certainly it demolished for ever the myth that the failures of the CAP could be blamed on the French as a permanent scapegoat for British criticism.

In actual fact, France was bound to make the running in the early years of the CAP simply because it was the largest country with the most farm land, and soon developed an interest in the need to export. These very facts, moreover, ensured that the French farmers had more political influence than most, and so had the greatest impact on policy. To tell the truth, British farmers became glad of their influence as the NFU lost ground at Westminster. The most acute moments of friction were more effectively dealt with farmer to farmer, which could break through tactical considerations in favour of the longer-term strategy. It was also appropriate, perhaps, that France was a nation which cared about its food and the quality of its cooking.

In that sense we have taken a leaf out of the French book in current developments, and may even be starting to abandon our traditional fixation with cheap food. Certainly we are finding much common ground. The Ferme de Vimer in Normandy, established by the *Farmers Weekly* and now tenanted by the Greens, has been a mine of information on what it's like to farm in France. So much so that quite a few British farmers have followed that example, and successfully settled on French farms.

Lord Plumb, himself, in becoming president of the European Parliament, led grass-roots British farmers into a better understanding of European agriculture, and the growing links being established with other parts of the world. Here again French farmers were in the vanguard with the foundation of AFDI (French Farmers and International Development). Having sent grain to Africa in response to famine in the 1970s, French farmers followed up with visits to see what was happening on the ground. They pursued the sound principle of looking and listening, only raising initiatives in response to needs and requests. Very soon links had multiplied between different French Departments and French-speaking African farmers, which were grouped under the AFDI umbrella. Many of these not only met immediate needs, but also established a wider understanding.

It paralleled the official work pursued at Parliamentary level through the Lomé Convention, which has developed an ongoing relationship between the EU and the ACP (Africa, Caribbean, Pacific) countries. With the advent of the World Trade Organisation, these arrangements are having to be adapted to a multilateral basis. But there is no question of the value they have contributed, both in terms of practical development, and the growing relationships between Europe and Africa. All of which will be greatly needed to ensure that the WTO takes due account of the many issues in agriculture, which cannot be assessed simply in monetary terms. This will involve questions which the WTO has not, up to now, considered part of its remit. But they cannot be ignored.

In considering the present day, it is important to find a vision which all twenty-five countries of the enlarged Union can make their own. According to *Farmers Weekly*, the ten new entrants have on average seen a 50 per cent rise in farm incomes in their first year. It is to be hoped that this is true, and that the improvement can be sustained even if not at the same level. In an arrangement which saw them begin full membership on a fraction of the support enjoyed by the existing fifteen members, it is important that there should be visible progress. Overall farm incomes for the same period in the fifteen were generally static, though in the UK they were reported down by 8 per cent.

While the immediate pressures of the Doha Round have to be faced, it is to be hoped that further evaluation of the CAP will not be hurried. There is need for the enlarged Union to settle down, and to consider its role and aims in the current context. So it would be a mistake for those with a clear agenda, however cherished, to push for the fulfilment of preconceived obligations. The trade benefits are by no means as great as are often stated. A study in Dublin University in 1985 ran a series of scenarios on the effect of a total abolition of EU farm subsidies. Overall they consistently showed positive results for Latin America, and negative results for Asia and Africa.

The tendency to quote the amount of money spent on EU cows is also totally misdirected, since it is not paid to dairy farmers but to processors. It is of course targeted at exporting the surplus milk produced, but the obvious harm done to developing countries will be eliminated, if the subsidised export of cheese, butter and skim milk powder is ended. Once support no longer interferes with trade, its abolition will do nothing at all to help farmers in developing countries. But it is also seriously mistaken to suppose that any extra export opportunities which may arise from freer trade will be sure to benefit African or Asian farmers. Most of such trade is in the hands of large companies, and the lot of those who harvest and pack the produce is often only slowly improved by strenuous pressure from unions and

NGOs. There may be wealth creation, but comparatively little of it is going to the alleviation of poverty.

The real opportunity for the ordinary family farmer lies much more in intra-regional trade, and sound policies for domestic agriculture. In 2004 there was only $2 billion of intra-regional trade in Africa, against a $50 billion demand for food crops, which is expected to double in the next fifteen years. So there is a huge domestic demand waiting to be satisfied. By the same token, as mentioned in the last chapter, 89 per cent of the EU's agricultural trade is intra-regional, and exports outside the EU are declining in response to policy decisions. At the same time the EU imports more agricultural products from the developing countries (€36 billion) than the US, Canada, Australia and Japan combined. This includes two thirds of all African exports worldwide. While the Everything But Arms Agreement (EBA) will grant free access to all exports from the 49 poorest countries (LDCs or Least Developed Countries).

In view of this it is fair to suggest that the CAP is moving in the right direction, and that in making radical changes it is important to take account of likely outcomes. Brazil has successfully challenged the European sugar regime at the World Trade Organisation. So changes are now coming under discussion (March 2005), and beet producers already understand that they will have to give ground to cane sugar. Sugar cane, although possessing a lower sugar percentage than beet, yields massively under irrigation, and so has the edge on competitive costs. It is one of the most productive plants in the world. But although sugar cane is grown in large and small areas, it is the big plantations which dominate the market. This means that Brazil and Australia would be most likely to take the bulk of any expansion in that market. It has been estimated in approaching the problem that Europe would have to reduce its sugar production by 2.8 million tons, with an accompanying reduction in the farm price for beet of 25–30 per cent, though higher figures have also been floated.

Some of its critics are aware of the early history of beet sugar development under Napoleon, and in the UK its beginning as late as the 1930s was given the thumbs down by a committee to examine the proposal. Finally, however, the minority report carried the day, to meet the security needs of the country in any future war. In the event that proved timely, but the UK remained a net importer of sugar until its entry into the EU, when its special arrangements were incorporated into the CAP. Not many people realise that sugar imported under these arrangements receives the European rather than the world price. So the LDC Sugar Group is proposing that reform should not be too hasty. They are suggesting a price reduction of 20 per cent, phased in over ten years rather than three. 'Managed access at remunerative prices generates far more development in the LDCs than unlimited access at uneconomic prices.' The Group estimates that such a period of stability would attract up to £265m in extra investment, generate £272m in export earnings and create 145,000 more jobs.

The position in the UK itself is that production at about 1.3 million tons per year is half the country's consumption. The other half comes from the ACP countries and Barbara Stocking, the director of Oxfam, has pointed out in a letter to the *Times* (25 November 2004) that unfettered liberalisation will harm the most vulnerable.

> Take sugar for example. Mozambique and Zambia are both low cost producers of sugar with long-term potential to be competitive. In these countries alone, trade in sugar could create 30,000 jobs given the right conditions. However if they are robbed of the chance to sell their sugar to Europe at higher than world prices they will not be able to afford to consolidate and expand

> the industry. Indefinite protectionism is not what we advocate, but some breathing space, a chance for poor countries to establish themselves, and arrive at a position where they can harness the potential of trade to reduce poverty.

Viewed from the angle of European farmers, sugar beet has played an important role in the balance of arable farming. Its place in the crop rotation has undoubtedly contributed to the maintenance of biodiversity and soil fertility. So in making a major readjustment, Europe faces what is likely to be a painful dilemma. New Zealand, on the other hand, made the sensible and generous long-term decision to continue getting its sugar from Fiji.

The possible balancing factor in the equation could be the production of ethanol from sugar as a contribution to cleaner energy. Ethanol can be blended with petrol to improve engine performance, but like other alternatives to fossil fuel it can only be economic if given favoured treatment through tax relief or other means. Clearly, however, the investment to pioneer such products should be borne by the richer nations, while developing economies have the chance to expand on basics. Brazil, having no domestic resources of oil, has however been investing in ethanol production for more than thirty years. It has made a great success of flexible-fuel cars, which can be run on both gasoline and ethanol in any proportion. In 2005 for the first time flex-fuel vehicle sales have overtaken those on straight gasoline. This was noted in a press release from the World Association of Beet and Cane Growers, which emphasised that access to markets without remunerative prices is of no value.

Currently the US is rapidly increasing its production of ethanol, mainly from maize. Production topped 35 million tons in 2004, up from 22 million the year before. France meanwhile is aiming to treble production of biofuels by 2007. This would involve

320,000 tons of ethanol and 480,000 tons of biodiesel (rapeseed oil). The UK lags woefully behind as the only home-grown biodiesel is made from waste cooking oil. Yet we are importing it from the EU, and, since the latest price rises for oil, seven million litres per month of ethanol from Brazil.

There is a catch here, however, in the failure of commodity markets to deliver reasonable prices. Pope Paul VI as far back as 1966 told FAO leaders and others concerned with the politics of food that they must set out to change the whole financial and economic set-up on a world basis. His words may have fallen on deaf ears, but the need remains, and it is time to take a fresh look at the possibilities of international agreement on a new way of doing business. The Fair Trade Foundation has raised the profile of the issue, and pointed out that it makes no sense to be paying prices close to or even below the cost of production. What is more, they have demonstrated that the public is ready to pay a bit more, if the extra will clearly reach the family farmer. Tim Bennett, the president of the NFU England and Wales, has already commented that 'we need to introduce fair trade for European farmers, as well as for bananas and coffee'. In this, he echoes the point made by the Foundation that an alternative model is needed, which puts sustainable development and social justice at the heart of international trade.

The Fair Trade Foundation, which was only set up in 1992, has made extraordinary progress in defining the practical requirements of such an alternative model. Today there are 370 certified producers' organisations spread over 45 countries in Africa, Asia and Latin America. They comprise 800,000 families in grassroots groups. The value of goods traded has been growing at 50 per cent per year since 1998, and in 2004 totalled £140 million in retail sales. For those participating, it guarantees a fair and stable price for their products, and extra income for estate workers to improve their lives. It also gives small farmers a stronger position in world markets, and creates closer links between producers and consumers. This helps to put the accent on production grounded in proper respect for the environment.

The Fair Trade minimum price is calculated to cover the costs of sustainable production and a sustainable livelihood. All stakeholders, including producers and traders, are consulted in the price setting process. There is an additional premium for investment in social, commercial or environmental development projects.

To take coffee as an example, the Fair Trade price for arabica coffee on the farm is 126 cents per pound. This comprises the Fair Trade minimum price of 121 cents per pound, plus the premium of 5 cents per pound. If the international price exceeds the Fair Trade price, then the Fair Trade price becomes the international price plus the 5 cents premium. In actual fact the international coffee price fell to 45 cents per pound in 2001, its lowest level ever in real terms. It slowly recovered to 75 cents in 2004, while the Fair Trade price remained stable. More details can be learned from the Fair Trade website – www.fairtrade.org.uk.

So far Fair Trade is a drop in the ocean, but it is part of a wider argument. We have to be sure that fresh opportunities for trade actually materialise for smaller countries and smaller farmers. Large-scale sugar production, as already mentioned, has traditionally come from large plantations with plentiful poorly paid labour. Now it is coming from the same plantations with large machines, a step forward for machine operators but a huge loss of jobs. Such changes are accepted as the law of economics, but the onward march of technology could be something we manage rather than absorb willy-nilly. Economists normally expect to explain the consequences of such developments, and it becomes easy to regard them as inevitable. But a more creative view is possible, when the accepted aim is to meet people's need for jobs and for a fulfilling life. The laws of economics depend heavily on an historical judgement of human nature, and attitudes can change. It could equally be the economist's task to devise systems which deliver a desired outcome through a change in motivation.

Competition is valuable, but it is not the only factor to be considered. The CAP can rightly pride itself on paying serious attention to the needs of rural society, and the destruction of that society can only be bad news for poor farmers struggling in developing countries. It is time to leave cheap food behind, and concentrate on the pursuit of excellence in quality and safety. It has not yet got through to the general public that, far from being comfortably insulated, farmers in both the rich and poor worlds are fighting to maintain their livelihoods. The table showing the decline of farm income in the last fifty years, quoted in the last chapter, makes that clear. Even large-scale operators proclaim a need for support.

In point of fact, economies of scale are easily exaggerated, and tail off much earlier than realised. Lord Haskins, when head of DEFRA's (Department of Environment, Food and Rural Affairs) Rural Delivery Review (May 2003), wrote,

> The greatest concern about huge supermarket chains is that their main drive for growth lies in getting even bigger. Eventually, as the Hanson organisation discovered, this formula fails. At some point Wal-Mart, like McDonalds, will run out of opportunities to expand.

It is then that people will be brought up against the need for improvements, which are not measured in bulk supply, especially if such supply depends on low wages or long hours.

Hopefully the CAP can contribute vital experience for the future, against all the odds. Could farmers shape it to become their own touchstone for future policy making? An arena where what they want for their own farms is part of a world picture. Where the spirit of solidarity growing in the West embraces the whole of Europe. That calls for an enlargement of heart and mind, and needs time for reflection and discussion. It cannot be hurried, and it needs to be spurred by the realisation that it is within our choice and our decision to lead the way. Yet such a reappraisal could lead to a change of direction worldwide, which weighs rural and urban society equally in the scales of future civilisation. It may be a far cry from much of the horse-trading which is associated with agreeing European policy, but the moral and spiritual considerations which uphold rural life are not in doubt for farmers themselves. They may be needed to play a more visible part in shaping the next steps, and accepting the responsibility for a world role.

Whatever the gap between vision and performance, we must at all costs hang on to the vision. To answer poverty and secure the environment through agriculture and forestry is a practical possibility, which requires a professional approach. Many political battles will have to be fought, but there is no doubt that this fresh direction would be a powerful uniting influence. It has not perhaps been seen as a European policy objective, but it undoubtedly makes sense. It could be what is needed to fill an apparent vacuum.

13

Europe – the enlarging union

Having asserted my conviction in the last chapter that the CAP is now aiming to fit in with world needs, it carries the germ of a vision which could see farmers contribute to the ending of world poverty and to sustaining the environment. So, given that the CAP is beginning to point the EU in the right direction, what can farmers contribute to the vision for an entire continent? That is an enterprise without precedent, even for those who took part in the reconciliation after the Second World War. Much of it is about further reconciliation and healing, both in the aftermath of Balkan conflict and the removal of the Iron Curtain in the East. The advent of ten new members into the EU in 2004 measures both the achievement and the need.

The kind of trauma suffered by countries of the Soviet Union over seventy years runs deeper than that of countries whose experience of communism was confined to the post-war years. But for all of them, individuals as well as nations, it was an experience which unhinged as well as destroyed many lives. Communist values were established in every department of life, and those who could best survive were those who could adjust to the system or found work which could be done with a low profile. I was truly surprised quite recently to find that economists of international experience could simply write off communism as a monument to the failure of central planning. It may be perfectly true, but it seems an utterly inadequate verdict on such a tremendous historical tragedy. Communism raised a great hope, and then wilfully destroyed it. For in death and destruction it created greater havoc than Hitler. Liberation is a just cause for celebration, but the foundations for freedom need much work. Yet thinking of communism as just an economic system may explain why many who hoped to be helpful were mainly concerned with spreading a free-market gospel which was utterly inadequate for the many practical and human problems revealed.

The fact is that for countries which retained a strong sense of their national identity a fresh understandng of national interest had to be worked out. Communism as an ideology had already been abandoned, but there had been little opportunity to develop alternative national goals beyond the immediate commitment to freedom. Communist government offered security of a sort, and under its wing jobs would be allocated to all, through the increasingly inefficient State Collectives. It might not offer a career, but it did offer the means to stay alive and fed until the structure disintegrated.

It may be significant that Poland, the largest of the new members of the Union, was also the most successful in preserving its own identity under communist rule. The national spirit of solidarity and the vital role of the Catholic Church combined to make an irresistible force. For farmers, Rural Solidarity was a part of this articulate patriotism, and succeeded in preserving 80 per cent of the old landholding structure of family farms. For those from outside who entered immediately after the opening of the country, it was almost as though the war had ended yesterday. Even farming methods were as though frozen in time, and many veterans on the land attested to a resistance, which was openly waged as late as 1956, and thereafter, although less visible, was anything but dead. One farmer we met told how, when he was arrested, his father, who would be left to look after the farm, had his arm wantonly and deliberately broken by the communist police. Another relished the story of how he had taken part in an action to liberate the prisoners in a nearby gaol.

All the same, once Solidarity had broken through, a united national objective was lacking. For a time the country was floundering to come to terms with the new circumstances. Rural Solidarity lost members, but branches which bucked the trend were those where local leaders continued to care for the practical needs of their members. It is almost surprising now to think of large collective farms lying idle for five years and more, while those who had worked there sought other outlets, and the buildings succumbed to neglect and vandalism. It took courage for those without substantial means to farm this land again, but the transformation is readily visible today.

The wartime settlement at Yalta, which had dashed Polish hopes, led unexpectedly to the germination of a new Polish/German relationship. Some of the Poles who came to the land ceded by Germany later found some sympathy with the predicament of East Germany. Unexpected links paved the way to positive developments when the Berlin Wall fell. In fact the stately home of the aristocratic von Moltke family, which had once been the scene for plotting an alternative to Hitler, became a centre for European reconciliation. Spontaneous friendships began to redeem the legacy of suffering and bitterness.

Driving through Saxony for the first time on the way to Poland, I caught the spirit of the tourist brochures which announced: 'Where an idea changed the world.' It led me to read a biography of Martin Luther to repair my historical ignorance, and to learn something of the significance of the Lutheran Church in the resistance to communism, and the change of attitude which finally brought down the Berlin Wall. From another angle, I also stood outside the Frauenkirche in February 2000 when the Duke of Kent handed over, as a gift from the UK, the seven metre golden cross, which was destined to top the restored church. It was a moving moment in the process of reconciliation, following the fire bombing of Dresden in the closing stages of the Second World War, while even as I write (February 2005) the final chapter has been reached, with the completion of this restoration. A Cross of Nails was the symbolic gift from the city of Coventry, itself flattened together with its cathedral in an early wartime blitz, which sealed this completion.

Against such a background, Saxony played host to a farmers' gathering, concerned to build bridges between East and West in Europe, and seek a wider purpose in the world. A Lutheran minister, concerned for rural development, regretted the loss of funds which forced the Church to cut back in this field. Some farmers, however, were ready not only to take on the technical challenges, but to look across frontiers in building a new future. Ekkehart Knoenagel, leader of a team of extension workers next to Poland's western frontier, exemplified this effort. He was determined to seek the welfare of all those farming in his district, and to create a new pattern in difficult circumstances. Although a meticulous planner, his spirit was equal to embracing teamwork with the Poles, for whom the best plan was often conceived on the spur of the moment, and who might turn up in double the numbers expected.

The Poles themselves are always generous hosts, and have that mixture of humour and stubbornness which carries them a long way in tackling new situations. Arriving forty years on, one was hardly conscious of the fact that Silesia had until so recently been German. Yet arriving at the frontier by car, the first sight to greet us was the queue of cars with trailers heading for Poland carrying the wrecks from road accidents in Germany. These could soon be resuscitated, and provide the first generation of vehicles on the road to freedom. The Poles are excellent mechanics.

Stanislaw Choma and Mishek Sroda are two of my friends who farm near Jelenia Gora, just across the frontier from Germany. Both have reclaimed ex-collective farms on hilly land, which is not easy going. Both are men who, whatever problems beset them, always have time for other people. It is a vital attribute in enabling them not only to reshape a new homeland, but to continue to care for those working on the land which they have left behind in the east. They feel a bond with the Ukraine, which can be the foundation for a relationship which urgently needs developing in the immediate future.

The name Ukraine translates as Borderland. One writer describes it as flat, fertile and fatally tempting to invaders. For the Russians it was part of Russia, and for the Poles part of Poland. Ukraine had never been an independent state until the collapse of the Soviet Union.

Whatever comes next, few can doubt that the election of President Yushchenko represents a turning point in the history of the Ukraine. It happened against the odds, and was buoyed by a powerful upsurge of national spirit, which brought people journeying from many parts to rally in Kiev. It was not just to elect a new president, but to set the country on a new path. The way ahead may still be tough, but a decision of the utmost significance has been taken.

The independent launching of an Association of young Civil Servants of the Ukraine, followed by a Centre for Political Education, reflect the serious efforts being made to establish a new democratic foundation. With an outreach created largely by young volunteers and a growing number of NGOs, an impressive contribution was made to the elections, and the training of the activists needed. It was an important addition to what became a total of 50,000 scrutineers and observers. The Centre for Political

Education has already had some links with Belarus and Moldova, so ideas will be developed in a wide perspective. An honest and ethical framework has been seen as the path to political development, and the great need now is to maintain the early enthusiasm.

The students of agriculture and the growing body of independent farmers will have a lot to contribute to this process, because agriculture remains an important part of the Ukrainian identity. Hundreds of students, who have come to the UK for fruit picking and casual farm work, have used their time well, not only in saving money but absorbing new techniques. While many of the pioneers, who have launched themselves as independent farmers, have succeeded in showing what is possible. Now could be the time to establish a professional identity, which would win the ear of policy makers, and command their respect as a national voice for agriculture.

For the farmers' contribution is cultural as well as political, and it holds the key to many of the controversial questions of environmental practice. Chernobyl has cast a doom-laden shadow over the kind of disaster which can overtake a whole region, but there are also vital questions to be grappled with in setting regular standards of good practice in the handling and use of chemicals, in soil conservation and in animal welfare. Even Stalin is credited with the planting of shelter belts to break the wind blowing across the steppes, but a creative agriculture today is marked by a whole range of possible innovations to suit a variety of climate and circumstance. Yet these will all be the fruit of a new and independent relationship with the land.

Already Black Sea grain is referred to as a potent influence on world prices, and fluctuations due to weather have been enormous. One English contributor to the *Farmers Weekly* (18 February 2005) was shaking his head over the urge to add value, when he sees himself as the producer of a commodity like wheat, and not a potential flour miller with all the investment and specialist expertise involved. One must sympathise with the need to make commodities profitable, but my friend Misha, in the Dnepropetrovsk province, has already set up machinery to make macaroni from durum wheat and to press oil from sunflowers. All the same, the pressure to stay afloat financially should not mask the need for a new deal worldwide. Following the EU's obligation to eliminate export subsidies, the Ukraine should be included in looking at ways to raise commodity prices to levels which bring a reasonable return. That should be a major preoccupation for the whole region. It is a position on which Europe could undoubtedly find a basis for unity, and perhaps help to lead the world into alternative thinking about the future.

Wherever this may be leading, one of the initiatives in the proposed European Constitution was the appointment of a foreign minister. It is a potent source of disagreement when it comes to taking positions on current geopolitical conflicts. Yet it could rapidly become the focus for Europe's mission in the world, and an expression of a common diplomatic purpose. Britain lacks conviction about the EU because people do not see the significance of this. It is taken for granted that there is no such thing as a European voice, even when people talk of the EU as a balancing influence on the United States. Such a desirable outcome would depend on renewing faith, and a popular engagement with the political process which has been sadly lacking. It is not something which can be conjured up by a PR campaign, but one anti-European MEP from Poland changed his view completely, just from seeing the European Parliament at work. So there is still something in the working out of the European ideal which can touch people at a deep level.

People are nervous of talking about Europe's Christian traditions in case it is considered exclusive, but that is not something which would have bothered Christ. Nor would he have had any hesitation in welcoming Muslims. His message was always inclusive of the whole world, but its challenge to absolute moral standards was uncompromising for the individual. That is a stumbling block for some, but the concept of an ethical dimension to foreign policy has widespread appeal. A new path, and the stamina to fight the battles to establish and sustain it, is needed. Certainly in the long run, nothing else can offer an inspired future for the UN. Democracy has to work within and between nations, so a regional identity can be the beginning.

It should be a source of pride to find we are entering uncharted territory, and moving from the past to new conceptions, where many assumptions about power and wealth may quite soon seem dated. Winston Churchill suggested that 'The empires of the future will be empires of the mind'. It was perhaps a recognition that ideas would be more important than military or economic might, and that mind and spirit would be the critical and sustaining power. European unity may not at present conjure up such thoughts, but it may be implicit in the many friendships which we form. Without such friendships supranational action would have been impossible. National interest can no longer be defined without reference to a vision of community.

Yet the question still remains, where will a new head of steam come from to carry the whole enterprise forward, and inspire young people to make it their life's work? Such an inspiration could reshape a world, where European unity contributes to a wider purpose. Perhaps it will help to light up our understanding of the United States, where sheer size and ethnic diversity still has to be assimilated into an all-American picture. It is part of a human evolution which is under way globally, and is not given the priority it should have. We need a sense of security firmly built on trust, which is the only cohesive force to hold society together. In a world where refugees and immigration are often regarded as a major headache, it is time to set new standards of humanity. Bureaucracy is not good at making people welcome, but we should all make the attempt. We can all address the agenda of healing hates and hurts, which takes time but is hugely important. It is said that the head must rule the heart, but unless the two are working together a true vision for Europe is impossible, and public opinion will stay immobilised. To see the Union mainly as a means to economic success is certainly mistaken. In the long run the UN can only function properly in a world without superpowers. That is an evolution in which Europe could become equipped to lead the way.

14

Developing patterns in agriculture – Asia

Asia is a continent where population growth puts pressure on the area of land available to agriculture. The area per head is only a fraction of a hectare and falling, yet farm production in most cases is holding its own. The Green Revolution has run its course, and has lifted the whole standard of living in areas where it has been most effective. But the need now is for a similar forward impetus reaching into every country and every corner. Professor Swaminathan has called for an Evergreen Revolution, which could establish a continuous advance in productivity per unit of land and per unit of water, but without any ecological harm. He likes the description of the Roman farmer, Varro, who said, 'Sustainable agriculture involves increasing productivity in perpetuity.'

That means intensive farming, but not necessarily large-scale farming. It needs to provide jobs, and create happy farming families. It also needs to be founded on a complete farming system, with crops and livestock interacting. Much more attention is being given to a balanced diet, and that involves a variety of crops rather than relying only on the staple cereals. Much progress is being made, and those who prosper must see how to help those still stuck at the bottom. There are now apparently one million NGOs registered in India. That represents an immense variety of private initiatives with the potential to become more and more effective. It is a major resource in bringing new life to the country.

In a continent of large countries and growing populations, Thailand is not well known. Yet it is a considerable food producer and exports just over half of its total production. By percentage if not by volume, this puts it among the top of the league of food exporters. It is a leading exporter of rice, but nowadays the other staple export – seafood and particularly prawns – is expanding faster. There is also a substantial increase in vegetables and fruit.

In many ways Thailand is at a crossroads, with a great need to bring together good governance and farming policies. King Bhumibol Adulyadej, who has been on the throne for over 50 years, has been a major influence in helping the country to advance to its present position. He has launched a huge number of agricultural projects, and has taken a close interest in their planning and performance. They have been particularly designed to demonstrate stewardship of the environment, with good soil and water management, and marketing schemes to help farmers reach out beyond their own immediate locality.

Politically, the King has been a moderating influence. Over recent years the direct political power of the military has diminished, although many retired generals still enter the political arena. Prime Minister Thaksin Shinawatra is the first civilian prime minister to fulfil a complete term, and to win a second one. He is a representative of the successful businessmen, who have been multiplying on the back of entrepreneurial skills, and a considerable capacity for making deals. There is a mixture of solid achievement and dubious speculation. So it is a moment when corruption needs to be squeezed out of the political system, as democracy opens its doors to a wider range of leadership.

As a leading agricultural exporter, and a member of the Cairns Group, one might expect Thailand to be in the forefront of Free Trade. But a Free Trade agreement with China, signed in October 2003, provoked a headline in the English-language newspaper *The Nation*, 'Are we being swamped?' (8 March 2004), Chinese merchants were ready to move, but the Thai side was ill-prepared. Border trade figures still show that Thai exports have the edge, but the gap is narrowing and complaints are rising about the difficulties encountered by Thai goods. 'Thai exports go through the zero per cent import tariffs at the border only to end up facing provincial taxes and value added taxes deep inside the mainland as well as non-tariff barriers.' Farmers' leaders took the Thai government to task for rushing the deal with China, and leaving Thai farmers and traders fighting rearguard battles.

At the same time, bird 'flu remains a lurking threat, though the spotlight recently seems to have shifted to Vietnam. More recently, it has reached Europe, where many experts fear the worst. But Thailand is recognised to have done a good job in containing the spread in poultry. Large-scale poultry and pig production shows that the Thais have embraced factory farming, with some highly successful enterprises. But perhaps they have not yet caught up with the questions being posed in Europe on animal welfare, and the modification these impose on production line techniques.

The real heart of Thai farming, however, lies with the small and medium-sized farms which are the backbone of crop production. They have suffered badly from the advocacy of chemical herbicides and pesticides, distributed without adequate training or a regulating body. But, helped in many cases by the Extension services, a quiet revolution may have begun. It found expression in a recent International Farmers' Dialogue organised by Chiang Mai University and the National Science and Technology Development Agency: Northern Network in November 2004.

At this dialogue, Dr Chamaiporn Tanomsridejchai, from the Department of Agricultural Extension in Bangkok, outlined some of the rethinking going on. She emphasised the need to encourage farmers' initiatives, and to stem the flow of educated people leaving the land. Extension workers needed to enter into the farmer's situation, and offer forward plans for the progress needed. This message was already being put into practice by the Faculty of Agriculture at Chiang Mai. They have done a great deal to promote co-operation in the region, and to spread the message of research in a personal way. Ajan Puntipa Pongpiachan, the animal nutritionist who initiated the Dialogue, has always sought to know the family background of her students. She aims to inspire her students with what they can do for their community and country, as well as how to carve out a professional career.

Already a number of farming groups in the Chiang Mai region are putting new life into their communities. Mostly described as organic, they honour the experience of the past, but also research new developments. Some are not against the use of fertilisers, but all look for some input from livestock – whether chickens, ducks or pigs. Chomchuan Boonrahong, of the Institute of Sustainable Farming Communities, comments on how attempts at high-input farming created debt, because farmers were unable to control the market mechanism. Now they are developing direct selling through farmers' markets, and at one I visited on the edge of Chiang Mai prices were commonly twice the local average. This was down to the trust built between buyer and seller, and the confidence of the consumer in the quality offered.

In the hill country further north, Nam Ru village is an illustration of what can happen in the space of five years. Helped by the International Heifer Project (Thailand), the financial core of development has been in pig production. (Although founded on cattle breeding the International Heifer Project now embraces all livestock.) Mitsamphan hybrid pigs have been provided, and the production of home-mixed food has been developed. While training in animal health has been provided in conjunction with the Veterinary Department at Chiang Mai University. The twenty-eight families in the village have learned to work as a team, and also have programmes in producing biogas, cultivating vegetables and water management.

The fruit of this work can be seen in the higher income generated – from 2000 Baht to 6000–10,000 Baht per month (65 Baht = £1 approx). This means that the children are able to have further education, and there is money to improve housing and even buy more land for agriculture. The financing has been through a revolving fund, to which participating members paid 15 per cent interest per year. Nine per cent of the interest went to the group's capital build-up fund, and 6 per cent to the Heifer Project's 'Passing on the Gift fund' for other needy families. The group began with 52 members paying 10 Baht each per month, and in 2004 there were 80 members with a total fund of 128,329 Baht. Already they are very proud to be able to donate for others, as well as continue their own progress.

Another group in the Chiang Mai neighbourhood grew from the experience of Khun Pipat. By 1986 his tobacco crop no longer went for export, and he had debts of 200,000 Baht. He decided as a first step to plant only what his family needed for themselves, and required all family members to write down their essential expenses. It was a case of starting again from the bottom up. From this strict self-discipline, he has built up a successful community of 200 farmers. They are organised in 8 groups, and through Integrated Pest Management (IPM) are moving towards organic farming.

With mixed cropping and animal manure from pigs and poultry, they now have a sustainable system. Some of their baby corn is even exported through one of the King's projects. Khun Pipat explains that his over-riding aim has been to develop people, and to reach out both locally and to the world. For him, standing on your own feet as independent farmers means honest dealing. He finds it a great reward that four or five graduates are returning to farm in his village, including his own son. A recent local survey of young people's finance has even suggested that those in the country were, on balance, better off than those in town.

All these experiences embody the beneficial side of Thai development where self-help, NGOs and Chiang Mai University combine to fulfil the King's vision of agriculture as a national vocation. Although it is vital that it should be financially sustainable, it is still a combination of the practical and the spiritual. So there is a network of understanding building up, which links experience in Thailand with Japan, Korea and Cambodia.

An interesting example of this cross-fertilisation is Rainbow Farm at Khilek near Chiang Mai. It was set up in May 2002 by Professor Shintaro Sugiyama of Japan and Tawan Hangsoongnem, who has a masters degree in agricultural extension from Chiang Mai University. Their first objective was to develop the techniques of organic farming in tropical areas, which are different from temperate requirements. Then, secondly, to spread these techniques in Thailand, and if possible to other tropical countries which may need them. In their view, this would mean giving a scientific basis to organic farming, which could be tested through helping small farmers to become self-supporting.

Against such a background it is interesting to consider some of the political and policy ramifications of these developments. Japanese agriculture is viewed by many economists as an example of totally unjustified support for rice growing. A Swiss businessman suggested to me that it was morally wrong for Japan to produce rice, when Thai farmers were looking for the chance to export. Yet surely rice growing is an essential part of Japanese culture, and the country would be a lot the poorer if it gave up on domestic agriculture in obedience to some economic law. I doubt very much whether the majority of Thai farmers I know would share the view that Japanese farmers should be put out of business to provide them with a larger market. There used to be a trading principle of willing buyer/willing seller, which aimed at mutual satisfaction. Coercion and litigation are poor tools to shape a new pattern of world trade. Japan is after all a huge importer of food, and Tokyo is one of the largest cities in the world. It would be suicidal to sacrifice their agriculture on the altar of Free Trade, and assert that they have no need of home production.

Cambodia was the best-represented country at the Farmers' Dialogue in Chiang Mai, and there were strong convictions from both Thai and Khmer to build on the friendships begun. History and politics may have made the relationship difficult, but the farmers are convinced they can help to heal the division. There is certainly a huge need.

The genocide made an almost fatal wound in the body of the Khmer people, but, against constant difficulty, there has been an heroic effort to restore and to rebuild. Orphans have been cared for, and farmers have begun to grow crops, though often without knowing whether they have legal title to their land. Above all Cambodia is a young country, with an average age of under 20 years, and young leaders are beginning to see that they can take responsibility for the future in a way they had not imagined possible. The signposts may be lacking, but a new spirit is helping to chart a fresh course.

NGOs also help to support the initiatives under way, and Cambodia has been quick to take up the System of Rice Intensification, known as SRI. This is a fresh approach to methods of rice cultivation, which has the advantage of raising yields without any added inputs. There are many details to be considered in the total practice, but the key to increased yields is the ability of the rice plant to tiller, and grow shoots which do not just die back, but produce good grain. To achieve this, a single plant is transplanted in the rows rather than a group of two or three plants. Less water is used at this stage, so that each plant can attain maximum root development. Excellent results are obtained with short-stemmed varieties, and a lower seed rate is needed. Long-stemmed varieties may be inclined to lodge (fall over). The system is now being used and evaluated in many countries, but in Cambodia the farms are extremely small and lend themselves to an intensive approach. It also means that a smaller area of rice can supply the family needs, leaving room perhaps for the cultivation of vegetables and fruit, and with small livestock to supply organic manure there is the foundation of a self-supporting system.

At the moment agriculture is not a favoured study in further education. In fact it tends to be a fall-back at university for those unable to gain a place in more popular subjects. Hopefully this will change as stability grows, and Cambodia not only meets its own needs, but becomes an exporter as it used to be. Certainly in next-door Laos the government is convinced about agriculture as the key to development, and many agricultural schools are being built. Exchange visits and the dissemination of knowledge are encouraged, but this does not yet extend to the formation of independent farmers' organisations, and the growth of a professional role in national policy. Vietnam is a neighbour making rapid strides economically, but the recent history of Vietnamese domination over Cambodian affairs makes this still a fraught relationship. Although here again there are early signs that the younger generation may be able to re-establish trust and friendship.

It was interesting to me that when an Indian friend came to the Chiang Mai Dialogue, he seemed surprised to find that he felt so much at home. The Indians have always been ubiquitous and active in other countries, but they do not seem so numerous in South-East Asia. Yet Indian agriculture has much to offer, including a powerful scientific establishment. My friend also felt that a farmer exchange would be mutually beneficial, and is keen to host some Thai farmers. He comes from an area in Maharashtra where the cultivation of seedless grapes has been developed over the last twenty-five years, and is now an important money earner for the local farmers. For many it has been the bridge from subsistence to commercial farming, and is described in Chapter 6.

That is perhaps the key to what India has to offer. Although opening to freer trade and beginning to break down excess regulation, India has always retained its own way of doing things. The long ethical tradition inspired by Mahatma Gandhi has ensured that adaptation to contemporary change is linked not just to technology, but to the deeper sense of human evolution. It is giving rise to the study of good governance and the ethical standards which sustain the most long-lasting businesses. Even more perhaps it is epitomised in listening to the inner voice, and the leading of the spirit which flows from it.

It reflects, among other things, the truth that development does not depend on catching up with the West or the degree to which people are herded into cities. It also makes possible the extension of the democratic way, where power is shared and all are aware of the need for eternal vigilance.

The IC Centre for Governance was set up in 2004 as an autonomous activity under the umbrella of Initiatives of Change (India). At its inaugural meeting Abid Hussain, a former ambassador to the United States, said that people were getting disillusioned with the working of democracy, but the solution was even more democracy. He raised the issues of the optimal size of government, the need for teamwork among the stakeholders in governance, networking and transparency at the operating levels by redesigning systems which were no longer relevant.

He also said that the centre could attempt to begin the gigantic task of reinventing the state machinery by sharing success stories widely across the country, and by sparking public debate on the role of ethics in good governance.

Mr Prabhat Kumar, a former Cabinet Secretary, comments that there is a palpable crisis of governance in many developing countries. This applied to India and China despite their enviable rates of economic growth over the last two decades. The world's leading 200 multinational corporations have a combined turnover of more than the GDP of all the developing countries. Yet not one is coming up with a design for rural water management, rural stand-alone power systems, housing and transport. There is not a single technology developed by a multinational for the villages of poor countries. If good governance can include the behaviour of multinationals, he is surely justified in pointing out such shortcomings, which arise from the pursuit of maximum profit as the priority.

At least in science there is recognition that, in farming, modern discoveries need to be married to the ancient wisdom of experience on the land, articulating the processes on which it is based. Organic farming has carried the flag of resistance to the methods of the multinational companies, where these have been limited to the commercial progress of their own enterprises. But science should be unearthing the truth for both, because it is committed to learning how nature works, and a proper understanding of the facts. That may sound strange, but facts need understanding, and the phenomenal growth of knowledge is forcing more consultation before we decide how to make proper use of it.

Professor Swaminathan's foundation at Chennai (mentioned in Chapter 4) has been particularly insistent on involving ordinary farmers in the development of new plant varieties and other fruits of research. He sees it as an essential part of moving from the laboratory to field trials. So it should not be a surprise to suppose that genetic science can make a useful contribution to organic farming. GMOs (genetically modified organisms) are not the only product of biotechnology, and tissue culture shows how disease-free stock can be established, even if it will eventually break down. There must surely be scope for producing plants with increased natural resistance to disease. Although the mutations which bring change need better understanding, particularly in the realm of bacteria, which mutate frequently. Applied science may continue to raise controversies, but pure science will always in the long run be a uniting force because it leads us to take an holistic view of nature.

The scientific world has always been an international one, and as such China is deeply involved. But it is not so easy to get to know Chinese farming, and I have not even begun as I have never been there. It is, however, possible to follow a little of the early beginnings of independent family farming, and to know that production has flourished where it has been stimulated by proximity to the great areas of economic growth, which have so captivated the world's attention. Although, for many of those in the heartland of China, the chief magnet of such growth has been to draw thousands from the farms in a search for jobs. All the same 310 million farmers with an average holding of half a hectare are battling to feed the 1.3 billion population, of whom some 500 million are in the cities.

Zhu Rongji when retiring as prime minister said that the priority need of national policy was to raise the income of the farmers. This showed recognition of the dissatisfaction in the countryside, where taxes and other charges, allied to local corruption, have aroused considerable hostility. Amid all the rightly publicised progress, there has been an increase in the relative poverty of farmers. From 2005, however, farmers will be exempt from local and national taxation, and will receive a subsidy of $1 towards the cost of seeds. New farmers' leaders are also emerging who have secondary school or better education, and tend to come from well-off families. It is to be hoped that they can be allowed a formal voice through independent farmers' associations. That would be an immense encouragement to the rural sector, and facilitate farmers' interaction with their colleagues in other countries. It would be a mistake to continue to plan for them, while excluding them from being social actors in their own right.

That is a mistake which has been all too common, and has been reinforced by the assumption that economies of scale are vitally important in agriculture. It is an idea which still holds sway in the West, where wages are high, and efficiency is in danger of becoming a mathematical calculation. But there is ample experience in Asia to show that land saving is more important than labour saving, and at the same time the productivity of labour can steadily rise. Japanese agricultural output increased within the existing small-scale farming system. In Taiwan also the knowledge, capabilities and learning capacity of small farmers were recognised and appreciated. Government supplied what the farmers needed in terms of research, advisory work and infrastructure. The result was a big rise in farmers' income. In Indonesia it has been a similar story. The increase in rice yield in Indonesia, which helped it move from being the world's biggest importer to self-sufficiency in 1985, would have been impossible without widespread technological change on small farms.

The story of dairy development in India is also widely known and authenticated. It transformed the country's milk marketing, and made its milk industry the largest in the world. Even the landless in India may own one or more cows or water buffalos, and from a local co-operative formed in the small town of Anand a huge business has developed. There are three tiers to this programme – village producers' co-operatives, district unions and regional federations, which receive technical and financial assistance from two semi-autonomous government agencies – the National Dairy Development Board (NDDB) and the India Dairy Corporation (IDC). The system is effective in ensuring regular milk pick-ups, prompt payment and accountability to the members. Veterinary services are also provided. Dr V. Kurien, who has been a main architect of these developments, has always seen himself and his staff as serving the farmers involved. The result is that poor farmers have acquired the organisational capacity to promote and defend their interests.

All of this confirms the vital role of the small farmers in Asia, and the speed with which they respond to new opportunities. In general, it is the small farms which combine crops and livestock. Larger farms tend to concentrate on crop production or livestock separately, and tend to be less productive per acre. An American company growing tomatoes in Thailand found it more economic to raise their tomatoes on contract with local growers than to use their own land with employed labour. It is important that these experiences are understood and appreciated, because it is clearly the way ahead for the immediate future. Population pressures are increasing, but so far the farmers are showing an impressive readiness to respond. Their potential and achievements should be fully appreciated.

15

Developing patterns in agriculture – Africa

At the time of writing (March 2005) Africa is moving firmly up the international agenda under the propulsion of Tony Blair's Commission for Africa, and the movement for debt relief. The Commission report aims to take on the bribe givers as well as the bribe takers, and perhaps help to improve public understanding of Africa's problems. One such misconception is that Africa must necessarily move from a predominantly rural continent to an urban one. The UN apparently calculates that within twenty-five years 400 million more, mainly poor, people will live in African cities. It predicts that urban slum populations will reach 332 million within 10 years. Yet agriculture could offer a road out of poverty which includes the generation of more jobs.

International aid for African agriculture fell by almost half in the last decade of the twentieth century. The focus now should be on small-scale farmers, who have the potential to generate and sustain economic growth and employment. Farm Africa has shown what NGOs can achieve when they work at village level to encourage farmer-led initiatives. Improved methods and training in crop and livestock production soon multiply. In the same decade of the 1990s, an improved strategy for milking goats in

East Africa has already reached 50,000 farmers and become self-replicating. African governments can invest in the infrastructure to expand such growth. In Kenya, smallholder farming generates 29 per cent of gross national product directly, and a further 30 per cent indirectly through processing and workers' wages.

One of these smallholders, Duncan Nduhiu, took part in the Chiang Mai Dialogue, described in the last chapter, and he was able to recount the progress made through the co-operation of local dairy farmers. He was one of the founding group which created the Nyala Dairy Multipurpose Co-operative Society. He tells how the conviction grew in him to call a meeting of local farmers and see how they could co-operate. His first attempt was unproductive, so he selected ten farmers to discuss the idea and they decided each would find ten more. In the event the group started with 210 members. One of them, George Kamau, told how his family and friends objected when he sold eight sheep to buy four 2000 shilling shares in the organisation. Their attitude changed a few years later, when extra shares and a dividend were distributed to members. What resulted from that first hesitant step has grown beyond Duncan's wildest dreams. Today he is organising training in animal nutrition and the use of organic fertilisers at six different centres, but the full picture of the Nyala Dairy is worth giving here. It is a story which reflects a reality which could clearly be multiplied, and already the number of participating dairy farmers has risen to 5000.

The full picture is set out on the website of the NGO Techno-serve under the headline 'Nyala Dairy Plant Transforms Entire Community in Rural Kenya, Improving Lives of Thousands'. Before 2001, business activity in the town of Ndaragwa had virtually ceased. The 10 small stands and one hardware store that lined the street of this small town in central Kenya were generating barely enough income to survive. Local residents depended on small, one- to three-acre dairy farms to earn a living.

Isolated by dilapidated roads in this dry, mountainous region, the farmers had no option but to sell their milk to travelling middlemen, who notoriously underpaid them. Frustrated by their situation, a group of over 2000 farmers founded the Nyala Dairy Multipurpose Co-operative Society.

The Nyala farmers' dream was to construct their own cooling plant so they could sell their milk directly to Kenya's large dairy processors and avoid middlemen. They turned to TechnoServe for help. In order to raise funds for the plant's construction and new equipment, TechnoServe helped the Co-operative to create a farmer-owned business. Nyala's 2700 members became dairy shareholders by contributing $28 each – not an easy task in a country where over 58 per cent of the population earn less than a dollar a day. But with this structure in place, financing was secured and in 2002 the plant was completed. On the first day of operation, 3000 litres of milk were collected.

With TechnoServe's guidance, Nyala Dairy's collection has since quadrupled to 12,000 litres per day supplied by 2100 small-scale farmers. The largest milk processor in Kenya, Brookeside Dairy Ltd, is now purchasing Nyala's milk, and because Nyala Dairy is a reliable, high-quality supplier, it can pass along higher prices to its smallholder farmers. Recently, Nyala paid $86,258 to rural producers – the highest monthly payment since its founding. The dairy also provides a store which offers affordable cattle feed, veterinary drugs and credit for its members.

In three short years, the dairy has already increased incomes for thousands of poor farmers, but what it has done for the town of Ndaragwa is truly extraordinary. Nyala Dairy has created a standard of living that none of the farmers ever dreamed possible.

What was once a ramshackle village on the side of a crumbling road is now a busy town centre that hosts speciality shops, a trading market, travelling businessmen and a growing population. Since Nyala Dairy's inception, more than 20 small businesses have been established for milk collection alone. The town also received electricity for the first time when the Nyala Dairy was established, and local entrepreneurs have opened several metal welding and automotive repair businesses. Some Ndaragwa residents even have cellphones – a luxury unheard of just a few years ago.

Before Nyala Dairy, local roads were in a constant state of disrepair and mostly unusable during the wet season – a major problem for farmers, who sell a perishable product. Now Nyala Dairy plays a lead role in lobbying authorities to repair roads, and has even undertaken road repairs itself. As a result, traders and others can now travel to Ndaragwa, which has encouraged even greater business growth.

The town's school system is starting to improve as well. In Kenya, secondary schools are not free and many children are forced to leave after primary school because their families cannot afford to pay school fees. Now many parents in Ndaragwa can send their children to school thanks to reliable income from Nyala Dairy. Parents have even started to contribute the money necessary to build new school dormitories. The Nyala Dairy is also having a beneficial effect on residents' health. The Ndaragwa health centre, which takes care of some 50 patients per day, confirms that the number of patients who cannot afford basic medical services and drugs has dropped by 50 per cent since the dairy plant opened. The number of nutritional deficiency cases among children is also declining, and more mothers are now bringing their children to the clinic for medical check-ups.

Women, who constitute 60 per cent of Nyala Dairy's active membership, are also becoming empowered by their increased incomes. After making household purchases, many are investing their milk payments into local projects established by women's groups, such as a home for Aids orphans. The orphanage has provided housing and food for more than 30 children and has already collected enough to pay the school fees for 20 of these children. They are in school today. 'The town of Ndaragwa is what it is because of Nyala Dairy. The town speaks for itself,' said Nyala Dairy chairman Samuel Ngure.

> Before the plant, people had nowhere to sell their milk. We were simply price takers. Now we are competing with other sellers. We are also reliable suppliers. When we give loans, we give favorable terms. Then people are able to pay for their children's school fees, purchase an extra cow or improve their homes. Now farmers are coming to us. I can proudly say that today we have a voice.

'The Nyala Dairy has come a long way, but there is still much work to be done,' said TechnoServe Dairy Business Advisor Fred Ogana. Nyala Dairy's next challenge is to maximise milk production, recently made easier by the installation of a new 10,000-litre cooling tank. Nyala's main customer, Brookeside Dairy, was so impressed by Nyala's achievements that it purchased the $40,000 tank for them in advance, under a three-year financing plan.

With 2100 farmers receiving higher incomes, that means about 10,000 family members are directly benefiting from this one business. If you include the workers that some of these dairy farmers can now afford to hire, and the new businesses that have sprung up around Nyala Dairy, the number is probably twice that. Ndaragwa now has great prospects for an even brighter future – and one that its residents are building for themselves (www technoserve.org).

It is to be hoped that the international initiatives being proposed will bring new resources in the areas of debt relief, increased aid, health campaigns such as Aids and malaria, and more transparent investment of capital. There is also much talk of increased trade, but the biggest opportunities for trade are intra-regional. It has already been pointed out that there is only $2 billion of intra-regional trade in Africa, when there is a $50 billion domestic demand for food crops, which is expected to double in the next fifteen years. External exports are around $18 billion, and while an expansion is desirable, more analysis is needed of who benefits, since the producer is at the bottom of the pile and commodity markets too often fail to deliver a reasonable profit. Where Fair Trade has been at work, coffee prices were maintained at $1750 per ton, when world prices fell from $1800 per ton to $1000. But in Kenya coffee production has been hard hit, where it used to be a regularly profitable crop.

Fortunately in tea there has been a happier outcome. The Kenya Tea Development Authority (KTDA), set up in 1964, has made possible the successful development of smallholder tea production. A pilot scheme paved the way for rapid expansion, and by 1988 there was an industry of 145,000 smallholders, providing green leaf to 39 factories operating without subsidy. This has helped to fulfil some of the ambitions of African farmers after the well-executed Land Settlement Schemes, which followed the Mau Mau uprising and independence. Jomo Kenyatta took a pragmatic attitude towards development, and made a considerable impression on the white farmers with the sincerity of his desire to see them play their part in an independent Kenya. He said there had been mistakes on both sides in the conflict, and they should move on without recrimination. All the same, he did later pose the question whether rising affluence could be reconciled with a moral foundation holding back corruption.

In neighbouring Tanzania the story was very different. President Nyerere's early vision of equality in a Marxist style, with collectivised villages, proved a financial disaster. Despite lip service to the importance of agriculture, the economic position undermined it. Between 1972 and 1986 the Tanzanian smallholder price for coffee, cotton and tobacco averaged less than half the world price. This reflected macroeconomic policies which failed to control inflation, and resulted in unrealistic exchange rates.

Since 1986 Tanzania has changed direction, and has every chance of establishing a healthier economic system. Kenya in contrast has stagnated, with growing bureaucracy and patronage robbing the economy of its dynamism. Both have need of better agricultural research, and more effective extension or advisory services for farmers. But there is no reason why both should not profit from networking together, and start to set a new pattern for East Africa, in conjunction with Uganda, another country on a new path.

Many Africans are taking up the cause of good governance in Africa. As Amina Dikede of Nigeria points out, 'Africa's future is in the hands of Africans themselves'. She and her colleagues in Initiatives of Change started with Clean Election campaigns in Kenya and Ghana, encouraging voters to refuse to take bribes or to support those who offered them. They are now mounting a Clean Africa Campaign for the whole continent. After meetings in the countryside in Ghana, Amina commented, 'That day showed me that country people are the conscience of the nation. They are ready allies. Their churches and mosques are full. If their leaders give them the message of Initiatives of Change, it will go to the nation.' Farmers in many parts of the world would echo that conviction, and long to make it an active reality.

Most are aware that it will be a long haul, but it begins with recognising the harm which civil wars and bad governance have done to African prospects. Fresh initiatives are vital to giving people the confidence to make their own opportunities, and to bring change to their countries. At an All-Africa Conference in Ghana (May 2003) President John Kufuor sent a strong message of support for these initiatives.

> I am also happy to learn that at this Conference a proposal to launch a Clean Africa Campaign will be on the table, and it is my fervent wish that positive decisions will be taken towards making this necessary and very

important crusade for a Clean Africa a reality. My government, and I am sure the people of Ghana, will be prepared to hold high the banner, starting with ourselves.

Following this initiative, Mrs Leonora Kyerematen became the head of the National Governance programme in the Ghana parliament. Speaking in Switzerland at the Initiatives of Change Conference in Caux (August 2004) she outlined the steps being taken.

The signs are propitious. Leaders are beginning to respect elections, recognising and accepting them as the only credible means of accession to power. All this was virtually an impossible phenomenon only a few years ago. The challenge is to foster the emerging consensus.

The strongest commitment to change for the better has been the adoption of the New Partnership for Africa (NEPAD) programme. As part of it the African Peer Review encourages member states of the African Union voluntarily to open up their systems of governance for scrutiny by their peers. African Union chairman, President Obasanjo of Nigeria, explained that it is intended that countries should correct themselves and do what is right in respect of governance and democratic practice. As at 8 July 2004 twenty-three countries had signed up to be peer-reviewed. Ghana is the first of three countries where the process is already under way.

For a continent whose leadership failures have been spectacular, the fresh start evident through the commitment to transparency to each other – and by extension to African people and the world – is deserving of international support

Ghana has enhanced the independence and credibility of the Electoral Commission, which has also kept faith with the people, minimising tension and improving trust. The government of Ghana also practises an open media policy, having repealed all restrictive media laws. The level of media freedom is probably without parallel in Sub-Saharan Africa.

To heal some of the social and political divisions, a Reconciliation Commission has been set up. These modest successes have not been a fluke. They are the fruit of principled leadership, especially the presidency, supported by the National Governance Programme, which I head, committed to a better life for all, and guided by a deep sense of personal morality. There is in place an Office of Accountability to check corruption in high places.

These are concrete commitments which represent a new beginning, and will encourage those who have doubted the rhetoric of a new deal. Africa has the heart power to give it reality, and perhaps to offer us an example of fresh honesty in politics, which is certainly not only needed in Africa.

It would be fair to say that making poverty history will require rather more than even the most optimistic outcome from a G8 meeting. That will certainly deliver valuable immediate help, but its lasting importance will depend on whether it signals a fundamental change in direction. The value of small farms and small businesses has been hopelessly underestimated, and it is assumed that large operations are the spearhead of the future. We become obsessed with figures and the billions which can be generated, when the greatest test may be how many million people can share in this new-found wealth. That is a concept of economic success which is not reflected in the rules of the Monopoly Board, but which is the fruit of full employment. Economics should be about sustaining livelihoods. What Africa will show the world remains to be seen, but it may well surprise us.

16

Developing patterns in agriculture – Latin America

It is thirty years since I visited Brazil, and I have not been again, even though I have often wanted to. Back then, although the military government was still in place, there was already a vision of the country as an agricultural giant. I was told that the farmers were the richest people in Brazil, and listened in Brasilia to a group of big operators joking about the difficulty for a rich man to enter the kingdom of God. Consciences were clearly touched by Jesus' comment in the Bible, and each was putting forward the other as the richest. It was said that 25 per cent of Brazilian land was suitable for arable cropping, and only 5 per cent was being used. Yet 80 per cent of exports were agricultural goods, and 90 per cent of the farm workforce was described as migrant. But there was no lack of practical men ready to seize an opportunity.

Today growth is once more rampant, though the finance minister, Dr Antonio Palocu, has been accused by some of administering an overdose of fiscal and monetary austerity. With prudent management and a boost from the 1999 devaluation, which has left the real at three to the dollar instead of parity, a record trade surplus of $30 billion is in prospect. Food now accounts for a more modest 40 per cent of exports, but is called the green anchor of the economy, while the 175 million cattle herd is already the world's largest.

Looking at particular cases, the state of Mato Grosso is one of the most eye-catching. Under current law 50 per cent of the land is potentially available for agricultural development, and so far only 30 per cent has been taken up. Caetano Polata, who began with 320 ha in the early 1980s, now has 25,000 ha. His main crops are cotton and soya, a large part of which are exported to China and Europe. The hub of his operation is an office in Rondonopolis, employing 45 people, while out on the farms there is a workforce of 420, which means about one person for every 60–70 ha. With good soil, an equable climate through the year and 60 inches of rainfall, it is no wonder that Caetano Polata describes it as a dream situation (*Farmers Weekly*, 10 March 2005).

One is reminded of pioneers elsewhere who in the first flush of their enthusiasm spoke of 'God's own country', but hardly stopped to consider what that might mean. The soils in Mato Grosso were at one time considered poor, being highly acidic with a Ph below 5. But lime and phosphate have transformed their performance, and with the organic matter accumulated through the years, they have the virtues of virgin territory. But in considering the long haul, that organic matter needs to be maintained and a truly sustainable farming system developed. This must include the farm workers who make up the rural community. They may or may not be adequately paid by current standards, but for the future they need access to higher education and the opportunity to take on further responsibilities as researchers or farmers in their own right.

Already in the late 1980s government attitudes were beginning to change. Jose Lutzenberger, the Brazilian environmentalist, recognised this, although he was disappointed with the slow rate of progress. It is now official policy to make ecological sanity prevail. The devastation of the forests reached a peak in 1987, but it has recently been revealed that 2004 saw a huge surge in illegal logging. It has provoked the government into action, and 89 people were arrested in a first sweep, which included the head of the environmental agency in Mato Grosso. The degraded areas offer immense opportunities for regeneration and sustainable farming. It represents an enormous area, almost the size of France, and much of it had been abandoned through the lack of capable people to redirect the existing agricultural research and extension services, and extend their range. As Lutzenberger wrote in 1992,

> Not only do we have millions of hectares of land degraded by bad agricultural and ranching practices, more being added every year, but we also have millions of uprooted people willing to work the land on their own or as farm workers, who have no access to it or no means to do so.

This brings up the central dilemma of land reform, which confronts a society where there is a huge gap between the rich and the poor. The MST, the Movement of the Landless or Sem Terra, has done a lot of effective work. It has had to battle on the margins of the law, but it has shown impressive purpose and discipline. As Father Arnildo has commented,

> The movement has gone much further than we ever imagined. There was always creativity, solidarity and the sharing of the few possessions people had. We in the CPT (the Pastoral Land Commission) dreamed of achieving agrarian reform, but not the development of a new man and a new woman. And this is what the MST is giving us today. I have been privileged to play a part in the process.

The MST's refusal to compromise has given it a kind of ideological purity, which hit a raw nerve in the Cardoso government.

Yet President Cardoso himself pointed out that what is most shocking is that hunger in Brazil is not the result of underdevelopment but of social injustice. The MST aims for a far-reaching transformation of society, though some wonder whether the younger generation will be seduced by TV and urban consumerism. But it need not be so. Some Marxist historians feel that the rural peasantry has become a spent political force which is largely irrelevant. Yet with the right encouragement their aspirations would meet overwhelming democratic endorsement. Brazilian territory is so huge and diverse that large and small farmers could co-exist without conflict. So it is vital that farming is seen as one industry, rather than a kind of dual track divided between commercial and subsistence.

Nor is Brazil the only country where such considerations are current. It is widely reflected throughout Latin American society, and revolutionary conflicts will have to give way to change by consent. But change must be embraced wholeheartedly rather than resisted. Already there are pockets of new life in agriculture in Chile, Uruguay, Paraguay and Bolivia. While Argentina, long an important player in farming, is rising again after financial disaster. My friend Jorge Molina, a soil scientist, used to say that his trials showed that soils on the Pampas could be maintained in equilibrium by four years of cropping being followed by four years in grass. But political difficulties have worked against the agricultural development which should have been possible. Indeed this has been a recurring theme in Latin America. Migrant farm labour has had to struggle to survive, and the odds have been stacked against it.

The migrants found their first champion in the US, when Cesar Chavez took up their cause. Born of Mexican-American parents who lost their small farm in the Depression, he became one of them, following the ripening crops up the West Coast. He founded a union for the workers in the sixties, and they decided to finance themselves with a monthly subscription of $3.50. The organisers, like Chavez himself, would live on a wage of $5 per week plus basic expenses. It was uphill work in the face of brutal intimidation, but they held fast to their commitment to non-violence.

Time and again Chavez rallied his people to continue the struggle. At one stage people across America responded to his call for a boycott on grapes, until the growers agreed to a union demand for a fair wage for the pickers. Through his leadership and the labour legislation he achieved, the migrant farm workers – though still poor – have rights and have hope. Breaking a fast in 1968, Chavez expressed the faith which inspired him. 'The truest act of courage, the strongest act of manliness is to sacrifice ourselves for others in a totally non-violent struggle for justice. To be a man is to suffer for others. God help us to be men.'

Some of these tensions are reflected in the recent history of farming in Mexico, particularly since 1970. The rapid expansion of agricultural output in Mexico between 1940 and 1970 made a vital contribution to overall economic development. Between 1950 and 1970 the population had risen from 26 million to 48 million, yet increases in wheat production not only met domestic needs but replaced imports. Agricultural exports peaked in 1972 at 68 per cent of Mexico's commodity exports (mainly minerals), and the major items were cotton and coffee. With the oil boom in the 1970s favourable macroeconomic policies were reversed. High inflation and drastic devaluation halted agricultural progress. Attempts at shaping policies to develop the small farm sector were thwarted by lack of finance, and the two-track structure was accentuated. Growth in production was negligible, and the inequality in income distribution meant unemployment and urban slums.

The net result of the economic collapse of the 1980s has been that progress has been concentrated on the large farms, which represent 12 per cent of Mexican farms but have 40 per cent of the arable land and 70 per cent of the machinery. Yet they provide only 20 per cent of farm jobs. There is great need for rural employment both on the farms and in associated activities. Public investment in research and infrastructure could provide the opportunities needed to lift productivity and raise farm incomes. This could also help to relieve the extreme population pressure on Mexico City, now over twenty million, where government policy has concentrated industrial development.

It is a mistake to suppose that a dual approach is needed on the grounds that large farms will meet commercial production objectives, while small ones serve only social and political ends. The Campesinos tended to be despised, and there was little confidence they could acquire the managerial and technical skills so evident in Asia. The result is that no grass-roots organisations of farmers have developed, or been encouraged as a matter of policy. While there is still scope for beneficial interaction between farming and the non-farm rural sector, it is now a much smaller part of the national workforce.

Costa Rica has also suffered from a divided two-track approach to farm development. In 1974 80 per cent of the money loaned to farmers by the banks went to large farmers. The small farmers, though much more numerous, only had a 20 per cent share of the capital on loan. Fortunately the small and medium-sized farmers are now represented by Upanacional, founded in 1980. This farmers' union recognises that agricultural markets need supply management, and that fair trade and fair prices should be the subject of international agreement. While in the domestic food market, the country has to have a policy of food security based on local production. It suggests that the simplest way to a new deal would be for the farmers themselves to outline a method of handling these international problems to their mutual satisfaction.

Carlos Solis, the general secretary of Upanacional, has recently (March 2005) organised meetings to seek improvements to the proposed Central American Free Trade Agreement (CAFTA) with the United States, and a national referendum has been proposed. Support for changes came from Larry Mitchell, chief executive of the American Corn Growers Association, whose outline of a different policy is given in Chapter 19. NAFTA – the North American Free Trade Agreement – has already shown a negative impact on Mexican agriculture. It is estimated that five or six million farmers have left the land because they cannot compete with cheaper US corn. The result is that huge numbers try to enter the US, legally or illegally, looking for work, shelter and just to survive.

In the battle to prevent such a thing happening in Costa Rica, Carlos Solis has pursued his search for the right way ahead in a thoroughly positive spirit. Any bitterness or reaction against the US has been laid aside in an effort to find an answer which includes everyone. It is an approach based on a careful research of the facts, and firmly linked to moral values. Jose Calvo, another leader of Upanacional, is a highly qualified agronomist who was educated in the US and has taught in several countries. Unusually he writes –

> Later I saw that I had to work on a small peasant farm in order to face up to all the problems the small farmer has to cope with. It was there that I learned we needed to organise to defend ourselves... We need to see things not just from outside: perspectives change greatly when you see things from within.

Both Carlos and Jose demonstrate that a militant spirit can still have a uniting influence in seeking radical solutions. This is an important quality in the effort to shape the future of Latin America. It paves the way for an inclusive democratic vision which has been lacking. It also accepts the responsibility of grass-roots farmers to contribute to policy making.

*   *   *

In considering the last three chapters, the circumstances across three continents may be hugely different, but there is no doubt of a common thread, linking farmers and farm workers to big opportunities for progress. Capital is needed, but new ways will grow through people rather than high finance. Future development does not lie in copying the West. Factory farming may already belong to the past, so far as massive concentrations of livestock are concerned. This is not to deny large-scale operations as such, but to assert that the overall aim must lie in creating the pattern of a new society. A rural community where health and education are twin pillars with agriculture in giving a new sort of 'green anchor' to the nations.

The future depends not on driving people into cities, but in giving them a destiny on the land. This is one of the messages from the developing world, which has yet to be heeded. A professional voice from the farmer will embrace technical expertise, co-operation and marketing, but it will also rest on the love of land and livestock, and high ethical standards. It will always be something cultural, beyond the measure of economic trends and statistics alone, but nonetheless financially successful.

How this will affect the course of globalisation remains to be seen. But the challenge to all those currently leading the way is to empower the people in a wholly new democratic wave. There is a great deal of creative energy waiting to be unloosed, and which is ready to grapple with the possibilities and problems of climate change. Globalisation concerns the countryside every bit as much as the cities, and there are signs that rural society is behind those who wish to rewrite the present script.

17

The twin poles of globalisation

The World Economic Forum at Davos and the World Social Forum at Porto Alegre are generally regarded as the opposite extremes in looking at issues related to globalisation. Yet in a discussion at the Conference Centre of Initiatives of Change at Caux, Switzerland, in July 2003, representatives from both sides declared a commitment towards a global civilisation which is more inclusive, more humane and more secure in its solidarity. Their differences lay in the way to achieve such objectives.

It seemed perhaps symbolic that the two Latin American presidents who attended the Porto Alegre meeting in 2004 were Lula da Silva of Brazil and Hugo Chavez of Venezuela. The Brazilian president, who has sought to give his country a sound financial base from which to reform, went on to Davos. The president of Venezuela, who is both popular and populist, went on to look at land reform settlements in Brazil. For some it may still be possible to build bridges, for others the alternative road is all that matters in confronting those who seem to be currently in control.

Sir Digby Jones, the chief executive of the CBI (Confederation of British Industry), lamented the lack of respect shown at Davos for manufacturing industry. Although his comments were mainly directed at the growing influence of the NGOs, it is a fact that the world has moved very rapidly from a 'capitalism of production' to a 'capitalism of financial flows'. The latter accord 70 per cent of capital investment to the developed rather than the developing world, and they tend to be directed by people who are mainly interested in good financial deals. Hence the unease at their growing influence is well founded.

At the discussion of 2003 in Caux, Ignacio Ramonet, cofounder of the World Social Forum, said he believed that economics had taken the place of religion and war as the guiding power at the heart of our civilisation. It is certainly true that the growth of materialism would be expected to bring about such a result, and that more emphasis on motivation would help us in changing direction. Despite the apparent proliferation of war, a strong majority would reject it as a way of settling any dispute, and faith in religion has foundered more on its image as an institution than what God can do in individual lives. Civilisation needs to be raised on great thinking and great living, and the two are intimately connected. What is most needed now is a coming together of different cultures in a common world enterprise to which the West is a contributor rather than a controller.

This would be a considerable change in the way globalisation is seen and practised at present. But the whole process is not something which should be driving us simply by its own momentum, and the fear that we may fail to compete. Someone else's knowledge cannot bind us simply because they have got there first. Nor is government's first duty to control people, but to serve and look after them. Those who seek control above all are clinging to an outdated view of power. Unhappily that may not be accepted for the present, either by supermarkets or superpowers.

It is, however, possible to question the contention that consumerism is a growing feature of contemporary culture, and that consumer power is driving globalisation. Considerable attention is paid to the axiom that the customer is always right. But when it comes to advertising, the weight of the effort is to persuade people what they ought to like, or to reinforce the desirability of familiar brands. It is the men with money who have the power and call the tune. Another question mark may be raised over how the cost-effectiveness of advertising can be measured, but it is clearly more often aimed at making sales than informing the customer. The pursuit of low prices as a primary aim, while obviously attractive to customers, tends to lower the profit margin and therefore calls for more volume to maintain profit. It is then mistakenly equated with efficiency. This may be hard to criticise in the current climate of competition, but it is a classic example of the failure to consider the whole picture.

A recent study of the food chain, and the 'food miles' involved from farm to shop, illustrate the fallacy of this. Professors Jules Pretty and Tim Lang have made an exhaustive study of the way food travels, and unearthed important data. Jules Pretty points out that, even if the amount of food flown annually into the UK doubles, it would still be 'peanuts' compared to the billions of tons of food carried around Britain by road. Every ton of food travels an average of 59 miles by road, whereas 20 years ago it was 46 miles. It is costing £2.4 billion, and clearly the bill is climbing rather than diminishing, and the report concludes that it is better to buy a local lettuce than an organic one from the other end of Europe. That reflects the conclusion that it is the environmental impact which is important. So that it is crucial to keep food as local as possible to the point of production, though it would be impossible to satisfy the twelve-mile ideal set out by the report's authors.

Tim Lang feels that the distance food has travelled should appear on supermarket labels. 'Supermarkets have invested billions in a hyper-efficient, just-in-time system of food distribution, and actually it's just cuckoo.' With huge out-of-town stores where acres of car park accommodate shoppers, the problem is then compounded, because shops are no longer local to the shopper. Tesco appears to be reversing this by buying up convenience stores, but it would seem this is only done to extend its influence when planning permission is becoming harder to get. It is not a belated recognition of the customer's needs.

Industrial food lends itself to the supermarkets' heavily centralised, highly mechanical distribution system, but fresh raw ingredients don't. Ready meals do not vary much between supermarkets, and offer a better margin than raw meat or vegetables. One meat supplier comments, 'They only make about 10 per cent in fresh meat, but they need 20 per cent to cover their costs... on processed meat products they can make as much as 43 per cent'. Certainly in our small town of Bromyard, there are three good butchers' shops, and there is no question that not only do they supply better value, but one can be sure their meat is local.

One former supermarket chief executive says, 'It's our job to prise every penny out of our suppliers. Every single penny comes either from them or from the customers. We try to do business ethically, but we also want to do what is best for our shareholders.' So buying tends to become obsessed by profit. One buyer, from the paper department, had replaced all the main suppliers and moved the chain's profit margin up from 25 per cent to 50 per cent. He said he intended to do the same with vegetables. Yet suppliers who complain or criticise are under threat of delisting (removal as a supplier) sometimes at incredibly short notice.

From 1995 to 2002, check-out prices were up 21 per cent, but farm suppliers' prices for main lines rose by 2 per cent. Bananas are apparently the best earner, and the breakdown per pound of cost is: plantation worker 2p; plantation owner 10p; international trader 31p (including 5p tariff); ripener/distributor 17p; retailer 40p (Blythman). In the recent Denby Pet-food chicken scam, when unfit chicken was passed off for human consumption, five supermarkets were involved as users. Professor Hugh Pennington – the expert on food poisoning – commented, 'I am surprised the supermarkets have been conned by criminals. There are clearly loopholes in the systems they operate.' Yet it could be simply the logic of looking for the lowest price.

More unsettling still is the evident intention of the bigger supermarkets to follow the example of Asda/Wal-mart in diversifying into other goods, which provide better margins than food. In 1999, before Wal-mart took over Asda, a basic pair of George jeans cost £14.99. In 2003 they cost just £4. The new format for superstores centres round the café hub – known as a 'dwell time' area – which reinforces the idea of shopping as a leisure activity. Customers can be hooked up to a main computer, and, if they like, can, with digital hand-held assistants, travel down electronic aisles on a screen. Trolleys are only used in a distribution warehouse, from which the shopping is delivered to a designated pick-up point. There are no checkouts either, the table console doubles up as a personalised checkout. The elements of this design are already being tested, but perhaps, when the first flush of novelty has passed, consumers may not be so sure that this is the way they want to go.

Certainly for quality and local specialities, the smaller supermarkets will begin to have the edge much more clearly. Even if we do not go so far as abandoning our cars for shopping, it may be time to rethink our strategy as shoppers. Contact with the goods one is buying is essential for fresh food, and the ultimate technology to deliver processed food should make us 'think values'. The result may be a turning point in seeing convenience as a chance to make real relationships, rather than enter a world of automatic provision. Most supermarkets now sell Fair Trade coffee, but the Co-op is the only one to switch its entire own-label range of instant and ground coffee to Fair Trade. It does so even though it has to pay two or three times more for its coffee, in order to commit to the principle (Chapter 19). It is clearly a plus point for co-operation over the dictation of the stock market. The pursuit of excellence may yet weigh in the balance against the pursuit of size, and win consumer support.

For, whether or not modern trends are entirely consumer-driven, it is undoubtedly true that only consumer decisions will change the present direction. It requires us to think more about the implications of our daily decisions. As Jules Pretty puts it, 'The most political act we do on a daily basis is eat, as our actions affect farms, landscapes and food businesses'. Buy British may have to be superseded by Buy Local, but the implications are essentially the same. Not that this needs to be a rigid rule. We export about 400 thousand tons of dairy products each year and import a similar quantity. French cheeses and other exotic items clearly account for much of this, and it is commendable to extend our choice. But it is important that it is our choice, and that we give it some thought.

It is a fair bet that not many people will lose much sleep over such matters, but they need to be made known all the same. Democracy is an imperfect road to improved performance, but it works best when there is transparency, and real motives are examined. It is the only arena where change may come from the spirit of the multitude rather than the established power. Such moments of collective action where the control of a few is superseded by popular insistence may be seen as rare historically, but less spectacular change is the daily bread and butter of democratic life.

All it takes is our daily involvement to create the opportunities, and to conceive that taking up world issues may shift our mental horizons. The World Social Forum at Porto Alegre in

2004 decided that for 2005 there should be concurrent meetings in all five continents. How these will be co-ordinated in shared themes and purpose remains to be seen. It is a considerable challenge. Yet it is surely a challenge worth taking up.

For those who have concluded that markets, not socialist ideals, now rule the world, a further dimension may be needed. The concept of the brotherhood of man still has a powerful attraction for millions, as the symbol of an inclusive society. Self-interest may hold good to a certain degree, but it will not tell you how faithful someone will be to their word. Interests can be compromised by many other considerations, and it is the commitment of a person's will and the disposition of their heart which will be crucial to the decision made. Trust needs to be rock solid, and to hold good in the extremes of adversity. It is a fundamental of human relations, rather than of economic relations.

Too many seek to explain globalisation solely in economic terms, which look to prices rather than to human values. Even talk of modernisation can be taken to mean promotion of a Western model, rather than a fresh response to the new truths which daily come our way. There can hardly be any meaningful globalisation process without truly global input, and that is what people instinctively feel to be lacking. The bandwagon is rolling in the name of wealth creation, but little heed is paid to the purposes for which that wealth is used. In the gospel of materialism, more money is good and that is the end of the story.

Farmers have always been as keen on the cash as the next man, but the lessons of nature also inculcate the importance of character. Endurance and persistence in adversity have always been part of the farming scene, and very few embark on a farming career feeling it will be an easy ride. Plenty are still passionate about farming, and believe the countryside can raise fine people. Their character is part and parcel, perhaps the most important part, of the crops we grow.

18

Growing character as a crop

> We need a revolution to carry the whole world forward fast to its next stage of human evolution – to outpace the growth of human power, wealth and skill with a growth in human character.
>
> Peter Howard

Character is formed in all sorts of ways, both consciously and unconsciously. It is a compliment for someone to be called a 'character', even if it implies eccentricity. It is the flowering of the individual personality, which marks out each one of us as different. It is also the bedrock on which human societies have been built.

J.P. Morgan, the well-known American banker and financier of the early twentieth century, always denied that his most critical decisions were mainly based on financial considerations. In one Congressional hearing the Committee's Counsel asked Morgan – 'Is not commercial credit based primarily on money or property?' Morgan replied – 'No sir. The first thing is character.'

Counsel – 'Before money or property?'

Morgan – 'Before money or property or anything else. Money cannot buy it.'

That sums up the basis of trust which should be at the heart of every great enterprise. We began life possessing the human nature with which we are endowed at birth, and we shall end it with the character which has resulted from our choices and decisions. Henry Drummond makes the point that to speak of God's glory equates with speaking of his character. As he puts it:

> Life is the cradle of eternity. As the man is to the animal in the slowness of his evolution after birth, so is the spiritual man to the natural man... What is glory? It is that of all unseen things, the most radiant, the most beautiful, the most divine, and that is character. Glory is character and nothing less, and it can be nothing more.

The Earth is full of the glory of God, because it is full of his character. The many wonders of nature on our planet reflect a Creator of infinite imagination. While the Universe itself dwarfs even our most grandiose conception of space, and brings us face to face with a concept of near-eternity. Not so long ago we could identify two galaxies in space, now scientists tell us we can identify two billion. It is a wonder in itself that they can tell us as much. I have never been more than a very ignorant observer of the stars in the sky, but apparently one of the furthest things you can see with the naked eye is in the galaxy of Andromeda, two million light years away. So what we see is not the galaxy as it is now, but as it was two million years ago. Our own sun is 93 million miles away and its light takes only eight minutes to reach us. It is a conception of distance which puts today's space research into perspective. All the same it was good to see the scientists clapping the safe arrival of their space probe on Titan (January 2005). It is a wonderful thing that we can enlarge our knowledge at such a distance, even if it is still only the first page in a huge encyclopaedia.

Some scientists are now speculating that the speed of development in the human brain seems to show that more than the ordinary processes of evolution must be at work. Whatever the truth of this, it is character rather than intellect which will decide the purposes to which our knowledge is dedicated. So it is in character that urgent growth is needed, and education is an integral part of it.

The education needed is not mainly in knowledge, but to equip us to grapple with the issues of life, and to maintain a continuous learning process. Sir Richard Livingstone, a former vice-chancellor of Oxford University, wrote in the 1940s of the evident moral confusion as the accepted signposts and fences of ethical behaviour along the road were increasingly obliterated. He quotes the twentieth-century philosopher, A.N. Whitehead. 'Moral education is impossible without an habitual vision of greatness,' and comments that no profounder statement has been made about education since Plato.

By the same token a job should be a vocation, something more than the sum of its everyday routines. Some jobs are declared too menial to be so elevated, but they needn't be. In the words of the old song of the Mississippi paddle steamers:

> Paint or tell a story
> Sing or shovel coal,
> You gotta get a glory
> Or the job lacks soul.

Although today vocational training is considered by some a second-class option beside higher education, job satisfaction does weigh with people as well as the money earned. The satisfaction of a job well done lies in its standard of excellence. Indeed, the government has moved recently to encourage vocational training for those not suited by the academic approach.

So many of the parables of Jesus have to do with farming themes, whether cultivating wheat and vines, or looking after sheep. This might seem inevitable given the farming society in which he lived, but all the same there are striking parallels between the ways of nature and the spiritual life. Nature is strong on continuity, and inculcates the qualities of determination, perseverance, respect for natural law and care in giving crops and livestock the opportunity to grow to their full potential. So the changes now taking place have to be viewed within that perspective. Brawn is giving way to brain in farming today, but it doesn't lessen the need to maintain the spiritual values which are at the heart of human life.

Looking back on my own path through life, I was passionate about farming, I didn't talk about it a great deal, but when I had the chance to work on a farm I stuck at any job which came up, including spreading muck by hand with a fork alone on a large field. I was also keen to take on any overtime which might be needed, and to prove my stamina. Physical strength certainly played an important part, though it had its downside, if carrying 100kg sacks contributed to the need for hip replacement in later life.

I also developed an early allergy to dust at threshing time. The storing of sheaves in stacks prior to the arrival of the combine harvester generates considerable quantities of spores and dust. After a few years, I couldn't last more than a single day on that job. So it is perhaps my good luck that there have been no later repercussions. The contractor who drove the combine which did our harvest for many years, before the days of cabs and air-conditioning, finally succumbed to a chest condition. His death is one of the saddest memories of my farming life, and is a reminder that one cannot simply rely on physical strength to plough through problems which may need thinking out.

On the surface the growth of mechanisation to make physical work easier has largely altered the balance of life. There are many more daily decisions to be made, and much greater need to think about the ethical implications of many of them. Yet it doesn't alter the moral imperatives which underpin business life

GROWING CHARACTER AS A CROP

and the development of new farming systems. My own special interest has been in livestock and grassland production, and the growth of genetic knowledge is now raising questions which are engaging the interest of the general public. Up to now there has always been room for improvement in yield and performance, and achieving that improvement has been immensely satisfying. But there is clearly a natural limit to the amount of milk one cow can produce. So the question arises, when is one pushing them beyond reasonable boundaries?

The answer must rely more on observation than academic calculation. If a cow has difficulty in walking when freshly calved and her udder hangs dangerously low, she is clearly approaching a physical limit. But while all high-yielding cows will have big udders, the cow with the biggest udder is not necessarily the highest-yielding. Having attained such high potential, the question also arises how the cow needs to be managed to fulfil it, and whether any problems on health can be attributed to the stress of production. There is probably an optimum level of production for any particular herd, and improvement lies in raising performance at the bottom end. The overall character of a herd has become more important than the presence of outstanding individuals, and modern genetics have increasingly begun to grapple with the problems of promoting overall progress rather than relying only on key statistics. There is already some hard evidence that very large herds of 4–500 and more perform less efficiently than smaller herds. While they are also likely to be more dependent on concentrated feed.

The latter point is relevant because, in the UK and temperate climates generally, grass and green forage is likely to be the most economic feed. Ruminant animals convert grass into milk and meat, and are thus less in competition with people for grain. Hence the fact that in the war UK livestock numbers were reduced, and grass was ploughed to grow grain. It could also be a safety valve for feeding a growing population if demand outruns supply. But more importantly livestock offer a way of restoring organic matter and fertility, which is likely to be less widely spread where cattle are concentrated in very large herds on a central site. Getting out and grazing grass is undoubtedly the cheapest and most natural way to produce milk.

All of which underlines the need for a farmer today not only to balance these issues with his experience, but to express them to an urban public. It calls for qualities of reflection and understanding, not only to preserve the heritage of the past, but to articulate new targets for today. The diversification projects added or built in to farming systems represent considerable creative effort. They have called for enterprise and risk taking. Where they have entailed direct interaction with the public, they have usually shown that trust is quickly established and the standards of country life readily appreciated.

For even in an age of instant communication local cultures can differ, though human nature remains the same everywhere. Character flowers in adversity, and life is meant to be a battle rather than a programme of endless welfare. Surprising then perhaps that, in an age of vaunted individuality and personal freedom, peer pressures still play such a big part in conformity whether for morality or immorality. Pride has always been at the centre of it for me, and it has been a long battle to become free of what other people think. I won some battles at boarding school where it first became an issue, but it was a long way short of winning the war. One of the key experiences proved to be with my eldest brother, Edward, when I went to manage our home farm for him.

He was eleven years older than me, so we had never been close. Then, when I went to manage the farm, he had the capital and I had the farm training. So I naturally wanted to impress him favourably, to secure investment in the development of the farm.

GROWING CHARACTER AS A CROP

It was a big lesson to find that his confidence in me grew when I was honest about mistakes, rather than saying everything was going fine. One incident particularly rammed this home, when two boars got out of their pens and had a fight. Although one of them was quite badly injured I knew he would recover, and felt there was no need to say anything to Edward in case he felt it showed bad management. My conscience was troubled by this, however, and the next day I decided to be honest with him. A small incident in a way, yet it was indicative of an attitude of dishonesty which I had allowed to develop, partly because I never articulated it even to myself. It paved the way to a new relationship as Edward and I subsequently became partners both in fact and in spirit.

Relationships are not built, however, just by observing the rules of good conduct. Absolute moral standards are the bedrock of character, and provide the soil in which it can grow. Honesty, Purity, Unselfishness and Love provide a realm to be explored as well as a code to be applied. It is the point at which we abandon enlightened self-interest for a more radical path. Certainly our current civilisation is not adequate for the challenges which it is meeting, and needs to appreciate the potential for human development. The essential climate for such an evolution of mankind is one of transformation rather than moderate improvement. It is the opportunity for ordinary people to take on a responsibility of which they may have scarcely dreamed. Yet if they have a complaint to make against how things are, that may be the starting point. Change in oneself can kick-start a whole process.

That is the democratic heritage which we increasingly have the opportunity to enter. It may not be enough simply to do our job within the confines we have accepted. For the farmer it is no longer enough to produce good food, important though that is. He must, in the current jargon, diversify into new fields such as taking on society at large and the needs of the nation. This will mean opening heart and mind not just to environmental questions, but to the issues of health, education and local society which may cross his path. At one time farmers were well represented in local government, but this seems to have been eroded by increasing concern to keep afloat financially, which keeps them on the farm and cuts off much human contact. If it is still increasingly happening, it is certainly not the way things are meant to be, particularly when questions of devolution are becoming more important.

Whatever the facts on this, however, the real point is to spark inner convictions into action. Farmers often become too easily isolated by following the path of duty and doing what their farm demands. But growth demands that a plant keeps pushing upward from its roots; or it may fall back and wilt. A tough character may survive, but will it be able to reach out to others?

For that is an essential part of a rounded character. It grows out of spiritual understanding, and it also brings understanding of other people. It cultivates the kind of selflessness which can make other people great, when so much of contemporary criticism is swayed by emotion, prejudice and self-will. Not only is it immature but it also contributes to the divisions in society, being incapable of looking beyond to a positive vision. Unless we are willing to help a person to overcome their faults, there is little value in pointing them out.

If growth is the only evidence of life, it also has a message for the spiritual life. Character is not just about conduct or altering our behaviour, it is about becoming a new person with new motives. It begins with the will rather than the intellect, but it depends finally on heart and will becoming one. It is not something we attain entirely by our own efforts, so although we can help it grow with the right cultivation, the outcome is in God's gift. It is genuinely like growing a crop, which may yield beyond what we have done to produce it, and beyond our expectations.

For its growth depends a great deal on what we give to others, and the sense of direction which commits us to a purpose in the world. Kipling expressed it in his ever popular poem 'If'. It is being true to what you believe that elevates human nature and brings it to the threshold of greatness. It is the hand to the plough in terms of endurance. It is the guarantee of a new world, and the only possible guarantee. Its outreach is what brings character to maturity.

Maturity comes from having a moral compass and using it daily to set a course. In these days of early retirement, it opens up new vistas of how the time can be spent. Supported by a pension and freed of the pressures to secure a living, it is a chance to develop new perspectives. It is a part of mobilising the lessons of life, and there is no need to feel left on the shelf. But when a friend told me that I could now begin to harvest the experience of the years, I didn't fully appreciate what he might mean. Now I have found some of that meaning in talking to and learning from farmers in many parts of the world, because it is an old saying that a farmer never really retires. The reality of the natural world and the weather in his face continue to hold him in thrall, and there is room to discover how much these things are common to farmers worldwide. Sharing experience, and seeking the spirit needed to shape the future, is rather different to offering advice on farm planning or day-to-day farming. But it is just as much needed, because it takes strength of character to envisage a destiny for mankind, and to make it a personal objective. That is what the Farmers Dialogue' is attempting, and its early efforts are explored in the next chapter.

19

The International Farmers' Dialogue

It is becoming quite popular to appeal to dialogue as the way forward, whether formal or informal. It suggests an exchange of ideas, where people actually listen to each other. On a personal level, friendship is its most natural expression. With friends we spontaneously share and develop ideas without any obligation or pressure to produce some agreed statement at the end. But in spite of this, such exchanges are formative and help to shape ideas which subsequently influence or change particular situations. This is clearly true of the network of committed people created through Dr Frank Buchman, which became known as Moral Re-Armament. Buchman's biographer, Garth Lean, records a comment from Cardinal König, Archbishop of Vienna till his death in 2004, after enquiring what Lean was writing. 'He was a turning point in the history of the modern world through his ideas.'

In response to Lean's request for his reasons for such a judgement, König wrote,

> In the last century, there was a feeling among intellectuals that we could build a better world without God. Then came the First World War, and many felt that many

things had gone wrong. Buchman was among them, and he began to think what could be done. His great idea was to show that the teaching of Jesus Christ is not just a private affair but has the great force to change the whole structure of the social orders of economics, of political ideas, if we combine the changing of structures with a change of heart. In that sense he opened a completely new approach to religion, to the teachings of Jesus Christ, and to the life of modern man.

König, who never met Buchman, based his assessment on his own observations in recent years.

> Wherever Moral Re-Armament is active there emerges a new world – in small circles first, but the activity shows how great the force is...If I consider the information which comes to me from all over the world, I see changes which are visible and social effects which are tangible. This must come from the faith of the man who was at the beginning, otherwise I could not explain what has happened since in so many places. 'By their fruits you shall know them.' From that fruit you go back to the root.

As Moral Re-Armament now becomes known as Initiatives of Change, those 'small circles' continue to multiply and spread out into different countries and cultures. Hopefully they will act as a yeast to raise a new spirit which can accomplish a silent transformation. Already there are a number of programmes to address particular situations, as in the Agenda for Reconciliation, which seeks to heal the conflicts and wars preventing positive action. There are many ways in which structures change, but a beneficial outcome depends heavily on the spirit which prevails.

The Farmers' Dialogue is just one of these programmes, and has grown from the sense that farmers needed to find a way to have active influence on current developments. It arose as a conscious movement of opinion, when Swedish farmers were faced with entry to the EU, and consulted with their French colleagues. This led to a first Farmers' Dialogue in Switzerland in January 1994. In the following decade, Dialogues have been held in Switzerland, USA, Poland, India, Cambodia and Thailand. In one sense, of course, dialogue is continuous, but even in local or regional gatherings there is often a benefit from the presence of some from outside the immediate neighbourhood. It helps in crossing the mental as well as the physical frontiers.

Just as Buchman always insisted that Moral Re-Armament needed to be different tomorrow from what it was today, there has to be continuous change. The Farmers' Dialogue is still in its early stages, but it is aiming to build up both a vision and an experience capable of creating a worthwhile future for our planet. It is answering poverty by helping farmers to build up economically viable and functional farms, which set out to make a positive contribution to the environment. This is the foundation which has been neglected, and farming families are the motor which drive it forward. We need to think less in terms of population statistics, and more in terms of functional families who seek health, education and creative work.

How much these aspirations are shared worldwide very soon becomes apparent, and the common conviction is that human relations are very much more important than technology. Understanding between farmers of different countries and climates is the first and easiest step. But all are aware that science and the food industry must be included in the consensus which is being sought. Too many of the goals identified can never be reached by the farmers acting alone, but if a coherent picture can be presented, they may become irresistible.

The problems of production are mostly bound up with technology, finance and education. But the key factor of profitability depends on price levels and marketing, which raise far more deeply interwoven issues. As I write (April 2005), a report has been issued in Washington which warns that almost two thirds of the natural systems which support life on Earth are being degraded by human pressure. It is the verdict of 1392 scientists from 95 countries and it is impossible to ignore such a weight of opinion, even if every detail may not be correct. A growing proportion of the world lives in cities, exploiting advanced technology. But the scientists warn that conservation of natural spaces is not just a luxury to be enjoyed at weekends. These are dangerous illusions which ignore the vast benefits of nature to the lives of the six billion people on the planet. We may have distanced ourselves from nature, but we rely completely on the services it delivers. All is not doom and gloom, but it is reassuring to find that when farmers get together there is plenty of positive experience to be shared.

As Assistant Professor Puntipa Pongpiachan said at the end of the Chiang Mai dialogue in Thailand,

> I am very impressed by the speeches given by farmers who have shared their experiences. The voice of a young Cambodian lady who asked for more Thai–Cambodian dialogue to improve the understanding between our two nations is a very good example of what has been started. I hope government officers together with politicians from both sides will use this opportunity to find solutions. This idea can lead to a solution of the conflict in southern Thailand.

She echoes the way such developments begin to give new hope for solving political divisions, which go far beyond the immediate day-to-day business of farm development.

Shailendra Mahato, from Jharkhand in eastern India, illustrated this when he told how he had ended a long-standing feud with another village leader, Thakurdas Mahato. Shailendra had long campaigned for an Adivasi state (Adivasi = Aboriginals) but he was a bitter as well as a passionate politician. When he decided to listen to his inner voice, two things immediately became clear.

One was to apologise to his wife for his neglect of home responsibilities on the excuse of endless political meetings. The other was to apologise to Thakurdas, the Congress Party leader and his great rival in the village, for his bitterness and hate. Despite the warnings of his friends against taking such a step, it soon resulted in huge progress in village development. Ponds were dug which allowed for fish farming and a certain amount of irrigation in a rain-fed, low-production area. Farm incomes rose steadily, and two new schools were built. Now the new state of Jharkhand has come into being, and Shailendra is dedicated to helping it become a pattern of this new spirit.

That is how the ripple of change in the village moves out into a wider arena, and begins to grapple with questions of national and international policy. These may not be solved so speedily, but a body of opinion has to be built up first to make it possible. To do that dialogue has to be constantly expanded, not so much to produce a blueprint as to make thinking in completely fresh terms acceptable.

Farmers do not have too much difficulty in accepting a spiritual foundation as the starting point. But it can clearly be a stumbling block in economics, where some will feel it not to be part of what is termed the 'real world'. What has to be faced right from the beginning is that any fresh consensus has to grow out of refining our motives, because that will decide what means are acceptable. We need the ability to put ourselves in the other person's shoes, and so begin to see the whole picture. Alternative thinking begins with those who wish to rewrite the script, but haven't a ready-made substitute. It is important to have an overarching aim, which in current terms must include the answer to poverty and the proper stewardship of our planet.

Economically speaking it is straightforward to say that farming must be profitable, and that conditions must be created to ensure that. But it is a great deal more difficult to deliver, particularly as it is not a sitting target but a moving one. Larry Mitchell, CEO of the American Corn Growers Association, draws a parallel with the difference between alleviating treatment and cure with a disease. Aids is a disease where drugs for treatment are more likely to be available to the rich than the poor. But while drugs alleviate the symptoms, they do not as yet provide a cure. In farming, the disease is low prices on the world market, and for the rich, subsidies are an alleviating treatment, a valued drug. But they will be more easily eliminated if a cure is found for persistently low prices. We need policies which improve prices worldwide, establish adequate food reserves and address production adjustments to enhance production of crops in short supply at the expense of crops in surplus.

Historically speaking, the United States has been regarded in the twentieth century as the breadbasket of the world, though that role is now passing to Brazil. Farm policy in the US has been powered by government investment in research, technology and extension services. Government has also assisted in arranging credit and marketing, but more recently it has been willing to intervene in the market place to stabilise prices and ensure farm income. In the last decade, however, the US government has abandoned market-stabilising tools in favour of trade liberalisation. In 1996 this policy shift was enshrined in the Freedom to Farm programme. Since then US crop exports have remained flat or declined, while farm incomes have fallen and government payments to farmers have rocketed.

The consequences of this policy shift have been global and catastrophic. Since 1996 world prices for the four chief exports of the US, maize, wheat, soybeans and cotton, have gone down by 40 per cent. The result has been lower incomes, hunger, desperation and migration. Europe, whether by luck or good judgement, has followed a different path. As described in Chapter 12, the Common Agricultural Policy is engaged in substituting a single farm payment for each individual farm, divorced from production but still having some control on its volume through milk quotas and land left uncropped (set-aside). Ann Tutwiler, when president of the International Policy Council on Agriculture, Food and Trade, commented that the US 'had queered its own pitch on agricultural trade, and Europe was now making the running'.

The real point, however, is to outline an overarching world policy on farming and trade. It is not about fighting our corner, but creating a consensus. The elimination of supply management tools in recent US farm legislation has led to record government payments of nearly $20 billion per year. Foreign competitors rightly charge the United States and Europe with 'dumping' excess production on world markets for less than the cost of production. This ratchets up the cost of competitors' farm programmes, and damages the agricultural economies of developing countries.

The Agricultural Policy Analysis Centre at the University of Tennessee (APAC) has already begun work on alternatives. It aims at policies which can increase market prices to reasonable and sustainable levels, and effectively manage excess capacity. Under the banner of changing direction to secure farmer livelihoods worldwide, they have run a computer model to test the possibilities. They have used a combination of three possible measures: 1) acreage (production capacity) diversion through short-term acreage set-asides and longer-term acreage reserves; 2) a farmer-owned food security reserve; 3) other price support mechanisms.

Using these tools, their model (for the US) showed the area of eight major crops dropping by 14 million acres in the first year. A price increase for major commodities of 23–30 per cent. A rise in net farm incomes, and a fall of $10 billion per year in government support. So they affirm that farmer prosperity in the US and the world is not only possible, but achievable.

Whatever the final form which might emerge, this is clearly a message of hope. It also helps us to envisage what may be involved in feeding a possible population of 10 billion in 2050. Broadly speaking the extra population will be in the developing countries, and the extra production will be a challenge, although a strong section of opinion considers it achievable. The adequacy of food supplies depends largely on national and local production, which emphasises the importance of small farm production in Africa. In Asia present production has stood up well, but there will be extra pressure on land area from the growth in population. Only in Latin America are there still substantial areas to be brought into cultivation. But whatever the pros and cons of what science may have to offer in new technology, it is clear that profitable farming will be an essential driving force.

In this scenario the present developed countries will be virtually static in population, and can therefore provide back-up in food production if needed. The reserves of land taken out of food cropping would be an obvious starting point, but there are also large quantities of grain being fed to livestock. It is estimated that 44 per cent of present cereal production goes for animal feed, and that three quarters of this is fed in the developed countries. A reduction in the quantity of meat eaten is clearly easily possible in case of need. It also underlines the desirability of feeding livestock such as cattle, sheep and goats on grass and forage crops rather than grain. So the fear that total world production might not be adequate is probably diminished, but there remain the uncertainties of climate change and beneficial action for the environment.

What will happen as the amount of carbon dioxide in the atmosphere rises is not a simple picture. The net result of both rising carbon dioxide and temperature is difficult to predict. Sometimes they reinforce one another, and sometimes they may cancel one another out. Moreover the short-term effects may differ from the longer-term ones, and individual species will react differently to them. Lloyd Evans, the New Zealand plant physiologist, points out that while predictions are basically unreliable, they do tend to show that the adverse impact of climate change will fall most heavily on the developing countries at low latitudes, particularly in West Asia, Sub-Saharan Africa and Central America. Agronomic adaptation will be possible, but he ends his book (*Feeding the Ten Billion*) with an assertion of the need for comprehensive teamwork.

> The further raising of crop yields to match further population growth without compromising the ability of future generations to meet their own needs, will require all the understanding, inventiveness and interaction of farmers, industrialists, agricultural scientists, educators, environmentalists, health care workers and policy makers which have brought us to where we are, and which have been sampled in this book.

Farming and biodiversity always involves striking a balance, but it is a balance which it should be quite possible to find. Now that the Common Agricultural Policy in Europe is moving towards payment for environmental benefits, there is no longer pressure for maximum production. In the UK 6000 miles of new hedges have been planted in the last ten years, and there is every sign that the general decline of many species of farmland birds has been halted. In Asia, area per head of the population has declined substantially, so there is no question of less intensive farming. Local self-sufficiency in rice is extremely important, as well under 5 per cent of the world rice crop is currently traded on the world market. Three crops of rice per year are possible with irrigation, and show no harmful effects, but agriculture's share of water for irrigation is falling. So any increase in irrigation will have to come from better water management. This is likely to mean smaller, local projects, which prevent seasonal run-off and add to the subterranean water by percolation. As Cornell University researchers have warned, 'Without careful pre-appraisal the farmers' promised banquet may turn out to be simply an engineering picnic.'

Meanwhile in the Western world energy crops may restrain some of the reckless consumption of fossil fuels. It is estimated that agriculture uses 5 per cent of the world's energy, but solar power provides vast quantities of energy through photosynthesis, which helps to keep a positive balance even when inputs rise. Growing oil seeds for bio-diesel or using wheat, maize or sugar cane for ethanol may not seem a good use of land when food is required. But if land is to be taken out of producing food, it is better that it should be used for an environment-friendly purpose. The rise in oil prices may this time be a permanent feature of the energy scene. The surge in consumption is, for the first time, outstripping the capacity of world refineries to deliver. So the consequent rise in price will begin to finance all sorts of alternative technologies. It is interesting to learn, for a start, that China now has the largest ethanol plant in the world. There are also a great variety of industrial crops being developed, some of which are only needed in small quantities. Others, however, may yet be in sufficient demand to require a substantial acreage.

That is the outline of the bigger picture which needs to be filled in, but there are also many local and regional questions which need to be settled along the way. As the Dialogue moves across the world, these can be given proper and detailed attention. While the presence of those from other continents will ensure that the solutions take account of the wider repercussions which may arise. Even the current crisis in Niger, caused by drought and locusts, would have been picked up sooner by more dialogue with and care for farmers on the ground.

As I have tried to make clear at the opening of the chapter, tackling human nature is an essential part of the whole process.

If we are passionate about Creation itself, and the development and sustainability of life, we have to consider the purpose of its Creator. That, of course, is not an exact study, but the growth of knowledge coupled with the lessons of history should help us to believe that there is an evolution of the spirit as well as of the body. They are not two different worlds, but one and the same. That perhaps is the final lesson which farmers learn from life on the land, and which they need to act on in facing the future. The international Dialogue is a means of sharing a new journey in this direction, which may provide the spirit to bring the previously unattainable within reach. It has certainly given me a new view of retirement, and the perspectives of advancing age. Extra time, if it happens to be granted, is invaluable for focusing on the essentials for change. There needs to be a concern for the future which does not seek control, and a determination which looks beyond a single lifetime. The power of the spirit may be the next lesson of evolution, and transform the terms of our global dialogue.

20

The unity of the spirit

Blessed are the pure in heart. They shall see God

St Matthew's Gospel, Ch. 5

We are all believers though we do not all believe in God. Many criticise St Paul for opinions real or imagined, which reflected the times in which he lived. Yet all accept the well-known letter to the Corinthians which speaks so eloquently of faith, hope and love. At the beginning of that letter he writes, 'God has qualified me to be a minister of the new covenant – a covenant not of written law but of spirit.' It is that covenant which may still offer us the chance of unity today.

Sir Jonathan Sacks, Chief Rabbi of the United Hebrew Congregation of the Commonwealth, explains that covenant means extended responsibility. Covenants are more foundational than contracts. Social covenants create societies, social contracts create states. Covenants are beginnings, acts of moral engagement. What we need now, he writes, is a 'covenant framing our shared vision for the future of humanity'. The core content may still be a subject for discussion and development, but a shared understanding will make us comfortable with our diversity and difference.

This is what makes it possible to conceive of a practical expression such as has been attempted in the UN. It succeeds better as a world forum than as a vehicle for action. But it can never go further than the member states allow. Yet it is universal in membership, and its charter is a symbolically important step in setting out a road for conciliation rather than confrontation. The founders created the General Assembly as a forum for discussion, the Security Council for keeping the peace and enforcing agreed standards, the specialised agencies to address transnational technical developments and problems, and the Office of Secretary General to run this vast machinery smoothly and efficiently. In March 2005, proposals for reform were published. In the words of the present Secretary General, Kofi Annan, 'It is for us to decide whether this moment of uncertainty presages wider conflicts, deepening inequality and erosion of the rule of law, or is used to renew our common institutions for peace, prosperity and human rights.'

In effect the UN has authority without power. When the US and her allies went to war in Iraq, they had power but lacked international authority. For some that was reason enough to hold back. For others the moral authority lay with those who had the conviction to act. Why wasn't the choice more clear-cut? The answer must be that moral authority is vested in those who live what they talk about, and most of us are aware of failures to measure up. Those who take responsibility put their authority on the line, and in the long run it is moral authority which will prevail in a democratic world because it can unite people. To give the UN that kind of authority may be one of the tasks of this century. As its name implies, its role is not world government, but co-operation between the nations.

As I have tried to make clear in the last chapter, it is this international teamwork which is shaping the Farmers' Dialogue, and looking forward to a common world policy for agricultural production and distribution. It is also what is required for understanding the effectiveness of environmental policies and for monitoring the local manifestations of climate change. It depends on trust more than competition, and that means producing character which is the fruit of a new motivation. It expects an evolution in human nature which will make this possible.

Such an evolution could meet many of our unsatisfied aspirations. Science, which has unlocked and explained to us so many wonderful processes, is in danger of being devalued in the public mind. Too many see it as dictating our options, rather than opening the door on a new world. But it is not the science which is distrusted so much as the scientists. The challenge of being human is to match our growth in knowledge with humility. If we accept the assertion that we are becoming to some extent in charge of our planet whether we like it or not, then we must never allow ourselves to believe that we are adequate for the task. So, whatever our ideals, we have to learn first to put them to work in our own lives. We shall never make poverty history by raising money, if it is not accompanied by a change in our own living and behaviour.

This will have to involve economic institutions as well as economic thinking, but the first step may be to recognise that there are better ways of doing business than those routinely followed today. State collectives have long been discredited, but it is far from clear that capitalist collectives are so much superior. They deliver profits, but do they have any other visible measure of their value to society? It could be that medium and small businesses are more truly responsive to the growth of democracy and widespread education, and so would guarantee a pluralist path. Certainly the belief in growth as an infallible indicator of economic health is due for reappraisal. As is the belief, that if everyone is getting rich, it doesn't matter if the gap between top and bottom gets larger.

This is the crunch point of the relationship between economics and politics. The assumption that, if the economy is performing, elections must favour the party in control will hopefully become less reliable. We shall learn to cope better with the mass of information and knowledge which is thrown at us, and form more balanced judgements. Economics needs to become an affair of the people rather than an affair of the stock market specialists and the opportunistic speculators. There has to be a commitment to right outcomes as well as financial success. Economists not only need to make viable plans, but to take responsibility for the situation in which they are involved. J.M. Keynes, the great economist of the twentieth century, resigned as consultant to the Versailles peace process in 1919 because of the inhumanity of the terms proposed. He told Lloyd George, 'I am slipping away from this scene of nightmare.' But a generation later he was still ready to serve at Bretton Woods.

Capitalism and its practices have to jump to a new level, if its evolution is to measure up to the forward march of mankind. At present the global invitation is simply to join our bandwagon on the road to prosperity. It will need to be more radical if we are really to make poverty history.

Only the spiritual evolution of mankind can offer a certain hope for the kind of change needed. Even our secular society will have to come to terms with that. It is the realm in which our spirits are stirred to take on great things, and to love our fellows. To be superhuman is the road to truth and to God, rather than to Superman or even Harry Potter. But we certainly need to seek such a path, though not so much in spectacular action as in the still, small voice.

When Paul writes of faith, hope and love, everyone agrees with him that the greatest of these is love. Why are we so sure? It is certainly true that, with all our growth in knowledge, it is the heart rather than the head which perceives priorities and defends the eternal values. We don't have to be intellectuals to understand the essentials in the modern world, nor do we need to shrink from making our voice heard.

The South African writer, Alan Paton, made the unusual observation that 'to love is to bring one's whole life under discipline'. Love is the first step in understanding human nature, but as he implies, it involves sacrifice, because it is also the essence of unselfishness. Such a sacrifice, however, is an affirmation of positive living rather than just self-denial. It means accepting people as they are, but always having a vision of what they can become. That is how unity is born, it is not something which has to be organised.

The values we share in Europe can also have the same unifying effect, and that unity can only be broken by deformities of the spirit, such as the desire to belittle colleagues to boost our own ideas, or to be more right than anyone else. To think in terms of what's right rather than who's right is a good beginning. Then it becomes clearer when temptation for national advantage has entered in, and when the effort to focus on a deeper purpose has been lost. A global purpose may become much more practical if we have achieved a continental vision, which is truly inclusive and does not depend on economic benefits. One of the good things about the EU, after all, is that it is bent on helping its poorer regions catch up with the rest. There is an acceptance of a shared destiny, which at its best transcends the argument over budgets.

When it comes to achieving the things we hope to see, it may be a long road. The point about putting a hand to the plough is not to look back. The ploughman strikes a straight furrow, and must continue till the task is finished. In one sense it is part of the work of eternity, yet it means fighting today's battles and creating new perspectives along the way. The farmer who commits to this way of life takes on a new priority. The home farm provides insights into nature, but the instant communication of knowledge and the growing range of science leads to the wider horizons of our planet, bringing changes both in practice and understanding.

Farming leads to a conjunction of the intimate and the global, the family and preoccupation with our place in the natural world. The nature of life itself still escapes us, but, whatever our philosophic bent, it is the Creator who shapes our ends and has made us part of a mighty Universe. For some it may be an impersonal force, though it is hard to reconcile that with the presence of love. The power of the spirit resides in the mysterious process known as character building, because in the end it is the people we become which will decide what happens next. Too often we only ponder the matter at funerals, appreciating the imprint our friends have left behind. But increasingly we can learn to celebrate in life the secrets of great living, which provide a bridge from doubt to certainty. It promises a century with faith reborn.

Bibliography

Beddington, Rosa, *Mill Hill Essays: 1997*, National Institute of Medical Research

Blythman, Joanna, *Shopped: the Shocking Power of British Supermarkets*, HarperCollins

Boobbyer, Brian, *Like a Cork out of a Bottle*, John Faber, Arundel

Branford, Sue, and Rocha, Jan, *Cutting the Wire: Brazil's Landless*, Latin America Bureau, London

Brown, Andrew, *The Darwin Wars*, Simon & Schuster

Calvo, Jose, *La Reconversion Agricola*, J. Calvo, Costa Rica

Campbell, Paul, and Howard, Peter, *The Strategy of St Paul*, Grosvenor Books

Drummond, Henry, *The Ascent of Man*, Hodder & Stoughton

Evans, Lloyd, *Feeding the Ten Billion*, Cambridge University Press

Fox-Keller, Evelyn, *The Century of the Gene*, Harvard University Press

Friedman, Tom, *The Lexus and the Olive Tree*, HarperCollins

Giddens, Anthony, *Summary of the Third Way*, Cambridge Polity Press

Gratry, Alphonse, *Les Sources*, Librairie P. Tequi, Paris

Gyatso, Tenzin, *Ancient Wisdom, Modern World*, Little, Brown & Co.

Hertz, Noreena, *The Silent Take-over*, William Heinemann

Howard, Peter, *Design for Dedication*, Henry Regnery, Chicago

Humphrys, John, *The Great Food Gamble*, Hodder & Stoughton

Johnston, Douglas, and Samson, Cynthia (eds), *Religion: the Missing Dimension of Statecraft*, Oxford University Press

Johnston, Bruce F., Kilby, Peter, and Tomich, Thomas P., *Transforming Agrarian Economies*, Cornell University Press

Kay, John, *The Truth about Markets*, Allen Lane

Keegan, William, *The Spectre of Capitalism*, Vintage
Korten, David, *When Corporations Rule the World*, Earthscan
Lang, Tim, and Pretty, Jules, *Journal of Food Policy*, vol. 30, March 2005
Lean, Garth, *Frank Buchman: a Life*, Constable
Livingstone, Richard, *Education for a World Adrift*, Oxford University Press
Mackenzie, A.R.K., *Faith in Diplomacy*, Caux Books
Porritt, Jonathon, *Playing Safe: Science and the Environment*, Thames & Hudson
Reid, Anna, *Borderland*, Orion Books
Sachedina, Abdulaziz, *The Islamic Roots of Democratic Pluralism*, Oxford University Press
Sacks, Jonathan, *The Dignity of Difference*, Continuum
Scase, Richard, *Living in the Corporate Zoo*, Capstone Publishing
Semler, Ricardo, *The Seven Day Weekend*, Century
Soros, George, *The Crisis of Global Capitalism*, Sage Publications Inc.
Stiglitz, Joseph, *Globalisation and its Discontents*, Allen Lane
Tolstoy, Leo, *A Confession and Other Writings*, Penguin Classics
Tudge, Colin, *So Shall We Reap*, Allen Lane
Turner, Adair, *Just Capital: the Liberal Economy*, Macmillan
Unwin, J.D., *Sex and Culture*, Oxford University Press

Index

ACP (Africa Caribbean Pacific) countries 120, 125
Action for Life 92–3
Addis Ababa 41
Adulyadej, King Bhumibol 139
Afghanistan 75
Agricultural Policy Analysis Centre, Tennessee 190
Ahmed, Nazimuddin 48
Al-Azhari, Dr Yusuf Omar 102
All Africa Conference of Churches 67
American Business Model (ABM) 59–60
American Corn Growers Association 189
Anand 149
Andromeda 181
Annan, Kofi 196
Arap Moi, Daniel 72
Argentina 166
Arnildo, Father 162

Asda/Wal-mart 172
Ashafa, Muhammad Nurayn 96–7
Asia Plateau 48–9
Association of Beet and Cane Growers 127
Australia 122

Bacon, Roger 22
Bangkok 143
Barnabas, St 99
Beddington, Dr Rosa 26–7, 34
Belarus 136
Bengal 114
Bennett, Tim 125
Berlin Wall 76, 131–2
Beuve-Méry, Hubert 66
Blair, Tony 16, 43, 69–70, 84, 115, 151
Blythman, Joanna 172
Bockmuehl, Prof. Klaus 98
Bolivia 163
Boonrahong, Chomchuan 141

Brasilia 160
Brazil 33, 44, 49, 55, 57, 63, 71, 122, 124, 160–3, 168, 189
Brent Spar 14
Bromyard 171
Brookeside Dairy 155
Brown, Andrew 10, 89
Brown, Gordon 68, 70
BSE 14
Buchman, Frank 99, 101, 184–6, 190–1
Buddhism 99
Bunyan, John 100

California 61, 75
Calvo, Jose 166
Cambodia 143–5, 186–7
Cambrian period 12
Cameroon 29
Campbell, Paul 98
Canada 50, 122
Cardoso, President 162–3
Catholic Church, Roman 84, 91, 132
Caux 158, 168–9
Central American Free Trade Agreement (CAFTA) 166
Century of the Gene, The 11
Chavez, Cesar 163–4
Chavez, President Hugo 168

Chernobyl 134
Chiang Mai University 140–5, 187
Chile 162
China, Peoples Republic of 54–5, 79, 99, 140, 148, 161, 193
Chinese Association for International Understanding (CAFIU) 79
Choma, Stanislaw 133
Christ Jesus 8, 83, 98, 138, 163, 182
Churchill, Winston 136
CJD *see* vCJD
Clean Election Campaigns 72, 157
Clinton, Bill 40
Common Agricultural Policy (CAP) 116, 123, 127–9, 189, 192
Commonwealth 81, 93
Communism 37, 76, 80, 96, 129–30
Congo, Republic of (Zaire) 62
Conservative Party 69
Cook, Robin 91
Cornell University 193
Costa Rica 165–6
Coventry 132
Cowper, William 94
Crisis of Global Capitalism, The 42–3
Dalai Lama Tenzin Gyatso 82, 86
Damascus 98, 105
Darwin Wars, The 10

INDEX

Darwin, Charles 4, 12, 64
Davos 168–9
Dawkins, Richard 27–8, 84
DEFRA (Department of Environment, Food and Rural Affairs, UK) 127
Denby Pet Food 172
Denny, Charles 53
Dickson, Revd K.A. 67
Diem, President 77
Dikedi, Amina 157
Dnepropetrovsk 135
Doha Round 121
Dolly (sheep) 25
Dresden 132
Drummond, Henry 5–6, 104, 176
Dublin University 121
Dumas, Alexandre (fils) 8
Dyson, James 54

Egan, Cardinal 77
Eosinophilia Myalgia Syndrome (EMS) 15
Ethanol 124
Ethiopia 41
Ethylidenebis, Tryptophan 16
European Union 41, 60, 62, 64, 69, 78, 81, 111, 117, 120–3, 129, 135, 199

Evans, Lloyd 197
Everything But Arms (EBA) 64, 122

Fair Trade Foundation 40, 50, 125–6, 156, 173
FAO (Food and Agriculture Organisation, UN) 125
Farm Africa (NGO) 61, 70, 151
FDA (Food and Drug Administration) 15
Feeding the Ten Billion 192
Fiji 124
Fischler, Franz 118
Foot and Mouth disease 14, 108
For A Change 97, 102
Fox-Keller, Evelyn 11–12
France 39, 42, 92, 119–20, 124
Frauenkirche 132
Freedom to Farm Programme 189
French Farmers and International Development (AFDI) 120
Friedman, Tom 37, 53, 89
Fukuyama, Francis 1

Gale, Mike 18
Gandhi, Kiran 48
Gandhi, Mahatma 7, 67, 146
Gandhi, Sonia 81

General Electric 44
Genome Project 11, 13
Germany 39, 60, 92, 132
Ghana 157–9
Ghandy, Sarosh 49–50
Giddens, Anthony 68
Gill, Sir Ben 31
GMO (genetically modified organism) 16–18, 27–30
Goldman Sachs 64
Gorbachev, President 77
Gratry, Alphonse 95–6
Green Revolution 138
Green, Tim and Chrissie 120
Greenpeace 14, 16
Greenspan, Alan 59
Griffin, Dr Harry 25

Halford, Nigel 18
Hammond, Sir John 22
Hangsoongnem, Tawan 143
Haskins, Lord 127
Hereford Times 33
Hertz, Noreena 39–40, 70
Hinduism 47, 50, 99
Hitler, Adolf 132
HIV/AIDS 89–90, 154
Howard, Peter 7, 98, 175
Hussain, Abid 146

Huxley, Aldous 100
Huxley, Julian 4, 11, 22

India 65, 81, 141, 146, 185
Indonesia 149
Initiatives of Change (IOC) 48, 74, 92–3, 146, 157–8, 168, 185
Institute of Sustainable Farm Communities 141–2
Integrated Pest Management (IPM) 142
Inter Faith Mediation Centre 96
International Farmers Dialogue 114, 140, 183, 184–194, 196
International Federation of Agricultural Producers (IFAP) 79, 119
International Heifer Project 141
International Policy Council on Agriculture, Food And Trade 190
International Monetary Fund (IMF) 62–4
Iraq 75, 78
Islam 47, 83, 90, 97–8

James, St 88
Jamshedpur 47
Japan 75, 99, 122, 143–4, 149

INDEX

Jelenia, Gora 133
Jharkhand 188
Jihad 90
John Innes Institute 18
John Paul II, Pope 91
Johnston, Douglas 92
Jones, Sir Digby 168
Jospin, Lionel 39
Just Capital 38

Kamau, George 152
Kamgono, Emmanuel 29
Kay, John 59–60
Keegan, William 66
Kent, Duke of 132
Kenya 72, 152, 156
Kenya Tea Development Authority (KTDA) 156
Kenyatta, President Jomo 156
Keynes, J.M. 38, 198
Kibaki, President 72
Kidder, Peabody & Co. 43
Kiev 133
Kipling, Rudyard 73, 188
Knoenagel, Ekkehart 132
König, Cardinal 184–5
Korea 143
Korten, David 43, 54
Kufuor, President John 157

Kumar, Prabhat 144, 147
Kurien, Dr V. 150
Kyerematen, Mrs Leonora 158

Lang, Prof. Tim 170–1
Laos 145
Le Monde 66
Lean, Garth 184
Lewis, C.S. 101
Lexus and the Olive Tree, The 37, 89
Lichtenstein, Conrad 28
Liu, Ren-Jou 93
Livingstone, Sir Richard 177
Lloyd George, David 198
Lomé Convention 120
Luther, Martin 131
Luttwak, Edward 91
Lutzenberger 161–2

Mackenzie, A.R.K. 75
Maharashtra 48, 61, 148
Mahato, Shailendra 187–8
Mahato, Thakurdas 187–8
Malaysia 54
Malta 81
Mandela, Nelson 36
Mao *see* Zedong, Mao
Marshall Plan 76
Marx, Karl 46, 59, 95–6

Mato Grosso 161–2
McDonalds 127
Meat and Livestock Commission 24
Melchett, Lord 16
Mexico 164–5
Minnesota Principles 53
Mitchell, Larry 166, 189
Mobutu, President 62
Moldova 134
Molina, Jorge 163
Monarch butterfly 28–9
Monsanto 13, 27, 35
Moral Re-Armament 7, 48–9, 74, 91, 99, 185
Morgan, J.P. 175
Mozambique 123
MST, movement of landless 162–3
Muhammad, Prophet 83, 90
Müller, Alfred 29
Museveni, President 89–90
Napoleon Bonaparte 123
National Dairy Development Board 150
Ndaragwa 152–5
Nduhiu, Duncan 152
Netherthorpe, Lord 119
New Partnership for Africa (NEPAD) 158

New Zealand 62, 124
Nigeria 93, 96–7, 103, 157–8
Nix, John 109
Normandy 113, 119
North American Free Trade Agreement (NAFTA) 166
Northern Ireland 84
Norway 14
Nyala Dairy 152–5
Nyerere, President 156

Obasanjo, President 158
Ogana, Fred 155
OPEC 51
Origin of Species, The 4, 6
Orthodox Church, Russian 82
Orwell, George 44, 100
Oxfam 123

Palocu, Dr Antonio 160
Pampas 163
Pandey, P.N. 49
Paraguay 163
Pasteur, Louis 28
Pastoral Land Commission 163
Paton, Alan 199
Paul, sixth Pope 125
Paul, St 87, 97, 195, 198
Peacock, Christie 79

INDEX

Penn, William 7
Pennington, Prof. Hugh 172
Pipat, Khun 142–3
Plumb, Lord 120
Poland 130–2, 135, 186–7
Polata, Caetano 161
Pollock, Prof. Chris 17
Pongpiachan, Ajan Puntipa 141, 187
Porritt, Sir Jonathon 16
Porto Alegre 168, 173
Prasad, V.N. 48
Precautionary Principle 13, 16
Pretty, Prof. Jules 170

Qu'ran 87, 90
Quakers 91

Ramonet, Ignacio 169
Religion – Missing Dimension of Statecraft 91
Robertson, Dr Ian 16, 31
Rongji, Zhu 62, 148
Ruhode, Tapiwa 31
Russia 133

Sachedina, Prof. 85
Sacks, Sir Jonathan 38, 195
San Francisco 75

Sassoon, Siegfried 101
Saxony 131–2
Scandinavia 42
Scase, Richard 41–2
Schuman Plan 91
Semler, Ricardo 44, 55–8, 71, 112
Sen, Amartya 5, 46
Sex and Culture 87
Shanghai 43
Shell, Royal Dutch 14
Shinawatra, Thaksin 139
Short, Clare 41
Showa, Denko K.K. 16
Siad Barre, General Mohammed 103
Silesia 132
Silva, President Lula da 168
Singh, Jaswant 49
Singh, V.P. 48
Smith, Adam 46
Smith, Arnold 91
Solidarity 130–1
Solis, Carlos 165–6
Somalia 103–4
Soros, George 42
Soviet Union 129, 133
Spectre of Capitalism, The 66
Sroda, Mishek 132
Stalin, Joseph 37, 134

Star Link corn 15
Stephen (martyr) 97
Stiglitz, Joseph 62–3
Stocking, Barbara 123
Strategy of St Paul, The 98
Streeter, Canon B.H. 4
Stude Rite Corporation 54
Sugiyama, Prof. Shintaro 143
Summers, Larry 42
Swaminthan, Prof. M.S. 19, 29, 84, 114, 138, 147
Switzerland 158, 168, 186
System of Rice Intensification (SRI) 144

Taiwan 94, 149
Tanomsridejchai, Dr Chamaiporn 142
Tanzania 156–7
Tata Company 47
Tata, J.R.D. 20
Tata, Jamshetji 47, 49
Technoserve 152–3
Ten Commandments 88
Tesco 171
Thailand 139–40, 144, 150, 186
Summary of the Third Way 68–70
Tiananmen Square 80
Tokyo 144

Tolstoy, Leo 83, 99
Truth about Markets, The 59
Tryptophan 15–6
tsunami 85, 94
Turner, Adair 38, 41–2
Turner, Graham 64
Tutwiler, Ann 190

Uganda 68, 70
Ukraine 133–4
United Kingdom (UK) 23, 39, 50–1, 69, 81, 91, 109–10, 118–19, 123, 125, 170, 192
United Nations (UN) 19–20, 47, 69, 74–5, 78, 103, 137, 151
Unwin 87
Upanacional 165–6
Uruguay 163
Uruguay Round 116
USA 7, 39, 51, 59–60, 63, 69, 76, 92, 124, 136, 146, 166, 186, 196
Vancouver 98
Varro 138
vCJD 14
Venezuela 168
Versailles 74, 198
Vietnam 77, 140, 145
Voluntary Service Overseas (VSO) 92

INDEX

von Moltke family 131
Wal-mart 127
Washington 186
Wells, H.G. 98
When Corporations Rule the World 43
Whitehead, A.N. 177
World Economic Forum 168
World Food Conference 1974 114
World Food Programme 32
World Social Forum 168–9, 173
World Trade Organisation (WTO) 50, 116, 120
Wuye, Revd James Movel 96–7
Yalta 131
Yugoslavia 78
Yushchenko, President 133
Zambia 123
Zedong, Mao 5, 37
Zimbabwe 31